Differential and Difference Equations

Leonard C. Maximon

Differential and Difference Equations

A Comparison of Methods of Solution

 Springer

Leonard C. Maximon
Department of Physics
The George Washington University
Washington, DC
USA

ISBN 978-3-319-29735-4   ISBN 978-3-319-29736-1   (eBook)
DOI 10.1007/978-3-319-29736-1

Library of Congress Control Number: 2016935611

Printed on acid-free paper

This Springer imprint is published by Springer Nature
The registered company is Springer International Publishing AG Switzerland

To the memory of my father,
Harry I. Maximon, who introduced me to the
wondrous world of numbers in the first years
of my life.

Preface

This work was originally conceived of as a consideration of difference equations, with the view of investigating the topics which are well-known in theoretical physics in connection with differential equations, among them methods for the solution of first and second order equations, asymptotic solutions, Green's function, generating functions, integral transforms, Sturm–Liouvile theory, and the classical functions of mathematical physics. The two subjects, difference equations and differential equations, are generally treated separately, with only a brief reference to the similarity of the respective analyses. However, as the investigation of difference equations proceeded, not only did the similarity become more and more evident but it seemed to provide a good way to make difference equations more understandable to those of us for whom differential equations are familiar tools of the trade. In presenting a given topic, the attempt has therefore been made, whenever possible, to follow the analysis for differential equations by the analogous analysis for difference equations.

It is obvious to anyone familiar with this subject that many topics have not been considered here. It was not our intention to write an encyclopedia and hence choices had to be made. Among the topics omitted are nonlinear differential and difference equations, partial differential and difference equations in two or more variables, and chaos theory.

Acknowledgment

A particular note of appreciation is due William C. Parke (The George Washington University) for his unreserved enthusiasm for this work and for furnishing the analysis which appears in Appendix A on the difference operator.

I also wish to express my appreciation to Joseph Foy (Arizona State University) for getting me over the first hurdle—the formulation of a letter proposing submission of the manuscript to a publisher.

Contents

Introduction

The history of difference equations, if one includes recursion relations, spans the history of mathematics.[1] It starts with the beginnings of mathematics in Mesopotamia of the second millennium BCE. There, recursion relations were used to calculate approximations to the square root of 2, although only one step of the recursion was considered, [17]. Recursion relations that may be regarded as a precursor of difference equations in that they go beyond the first step were presented in some detail by the very prolific mathematician and engineer, Heron of Alexandria (c. 10—90 CE). In *Metrica*, the book in which he describes how to calculate surfaces and volumes of diverse objects, Heron shows how to obtain a second approximation to the square root of a number that is not a perfect square, [21]. A further example in which recursion is presented explicitly is found in the book written in 1202 by Fibonacci (c. 1170—1250), *Liber Abaci* (Book of Calculation), concerning the growth of an idealized rabbit population, leading to the sequence which bears his name. At the present time, recursion relations introduce one of the most significant concepts in physics today: chaos [14].

In Chap. 1 we present formulas for the operators E and Δ frequently used in the analysis of difference equations. In Chap. 2 we consider the solution of homogeneous and inhomogeneous nth order differential and difference equations. Section 2.1 presents the method of variation of constants, also known as variation of parameters, which may be used to obtain the solution to an nth order inhomogeneous equation when n linearly independent solutions to the homogeneous equation are known. This is applied to both differential and difference equations (see subsections 2.1.1 and 2.1.2, respectively). We also formulate the matrix equivalent of all the equations, resulting in first-order differential and difference matrix equations. In Sect.2.2 we present the method of reduction of order. This transforms an nth order linear equation (homogeneous or inhomogeneous) into an equation of order $n - 1$ when one solution of the homogeneous equation is known. This is carried through for both differential and difference equations. We present these two methods—reduction of order and variation

---

[1] A succinct history is given in the book review by Kelley [28].

of parameters—separately, since that is how they are generally found in the literature. However, as has been shown in a succinct article by Phil Locke [31], each of these procedures can be viewed as a particular limiting case in the solution of an nth order linear inhomogeneous equation when m linearly independent solutions, $1 \leq m \leq n$, of the nth order homogeneous equation are known: $m = 1$ corresponds to reduction of order, $m = n$ corresponds to variation of parameters. Related treatments may be found in [18, Chapter IX, Sect. 3, pp. 319–322] and in [20, Chapter IV, Sect. 3, pp. 49–54]. In Chap. 3 we consider in more detail first-order differential and difference equations. In Chap. 4 we consider second-order equations, comparing the results obtained by the methods of reduction of order and variation of constants. In Chap. 5 we consider self-adjoint differential and difference equations. Chapter 6 deals with Green's function, for differential equations in Sect. 6.1 and for difference equations in Sect. 6.2. In Chap. 7 we consider generating functions, z-transforms, Laplace transforms, and the solution of linear differential and difference equations. Section 7.1 deals with Laplace transforms and the solution of linear differential equations with constant coefficients. Section 7.2 deals with generating functions and the solution of linear difference equations with constant coefficients. Section 7.3 deals with Laplace transforms and the solution of linear differential equations with polynomial coefficients. In Sect. 7.4 we present an alternative method for the solution of homogeneous linear differential equations with linear coefficients. In Sect. 7.5 we consider generating functions and the solution of linear difference equations with polynomial coefficients. Section 7.6 gives an extensive treatment of the solution of homogeneous linear difference equation with linear coefficients. Subsection 7.6.1 considers the solution of second order homogeneous differential equations with linear coefficients through transformation of dependent and independent variables. Subsection 7.6.2 considers the solution of second order homogeneous difference equations with linear coefficients through transformation of dependent and independent variables. Chapter 8 presents a dictionary of difference equations with polynomial coefficients. In Appendix A we derive an expression for the higher-order difference of the product of two functions. Appendix B deals with notation used in this work. Appendix C presents derivations of useful formulas dealing with the Wronskian determinant. Appendix D presents derivations of useful formulas dealing with the Casoratian determinant. In Appendix E we give a proof of Cramer's rule, which we employ throughout this work for the solution of matrix equations. Appendix F deals with Green's function and the superposition principle. In Appendix G we derive the inverse of a few generating functions and Laplace transforms that are of particular use in the solution of second order linear differential and difference equations with linear coefficients. In Appendix H we give a few of the transformations of the hypergeometric function $_2F_1(a, b; c; z)$ which have been useful in the analysis presented in this work. In Appendix I we list the confluent hypergeometric functions which result from different choices of integration path for the integrals given in Eqs. (7.200–7.203). In Appendix J we consider the second solution to the difference equation for the confluent hypergeometric function in which the usual first parameter is a non-positive integer and the second parameter is a positive integer, i.e., a degenerate case for which the usual second solution is no longer independent of the first. We derive a closed-form polynomial solution for which details are given in [37]. We show that a solution in the form of an infinite series may be obtained from the known expressions for the confluent hypergeometric function, as given in [36, Sect. 13.2(i), Eq. 13.2.9].

Chapter 1
Operators

In the expression for a differential equation the derivative operator $D \equiv d/dx$ and powers of this operator, D^k, $(k = 1, 2, \ldots)$, operate on a function $y(x)$. The rules for D^k operating on the product or the ratio of two functions are dealt with in elementary calculus. In this chapter we consider the corresponding operator, Δ, for difference equations and derive expressions for Δ^k operating on the product or the ratio of two functions and compare them to the corresponding expressions for the derivative operator.

There are two operators which are frequently used in the analysis of difference equations: E and Δ. Although we will not rely exclusively on them in our work, they often prove useful. They are defined by

$$E y_n = y_{n+1} \tag{1.1}$$

$$\Delta y_n = y_{n+1} - y_n \tag{1.2}$$

from which it follows that

$$\Delta y_n = (E - 1) y_n \tag{1.3}$$

or

$$\Delta = E - 1, \qquad E = 1 + \Delta \tag{1.4}$$

These operators may be applied successively:

$$E^2 y_n = E(E y_n) = y_{n+2} \tag{1.5}$$

or, more generally,

$$E^k y_n = y_{n+k} \tag{1.6}$$

© Springer International Publishing Switzerland 2016
L.C. Maximon, *Differential and Difference Equations*,
DOI 10.1007/978-3-319-29736-1_1

Similarly,

$$\Delta^2 y_n = \Delta(\Delta y_n) = y_{n+2} - 2y_{n+1} + y_n \tag{1.7}$$

and, more generally,

$$\Delta^k y_n = (E - 1)^k y_n$$

$$= \sum_{j=0}^{k} (-1)^{k-j} \binom{k}{j} E^j y_n$$

$$= (-1)^k \sum_{j=0}^{k} (-1)^j \binom{k}{j} y_{n+j} \tag{1.8}$$

Just as the above equation expresses $\Delta^k y_n$ in terms of a sum over y_{n+j}, it will also be useful in the analysis that follows to express y_{n+j} as a sum over $\Delta^k y_n$. From (1.6) and (1.4) we have

$$y_{n+k} = E^k y_n$$

$$= (1 + \Delta)^k y_n$$

$$= \sum_{j=0}^{k} \binom{k}{j} \Delta^j y_n \tag{1.9}$$

In analogy with many expressions for the differential operator, similar expressions exist for the difference operator:

$$\frac{d(u(x)v(x))}{dx} = u(x)v'(x) + u'(x)v(x)$$

$$\Delta(u_n v_n) = u_{n+1} \Delta v_n + v_n \Delta u_n$$

$$= u_n \Delta v_n + v_{n+1} \Delta u_n \tag{1.10}$$

Similarly,

$$\frac{d}{dx}\left(\frac{u(x)}{v(x)}\right) = \frac{v(x)u'(x) - u(x)v'(x)}{v(x)^2}$$

$$\Delta\left(\frac{u_n}{v_n}\right) = \frac{v_n \Delta u_n - u_n \Delta v_n}{v_n v_{n+1}} \tag{1.11}$$

It may be noted in (1.10) that the difference operator lacks the evident symmetry of the differential operator. This has the consequence that the higher-order difference of the product of two functions is more complicated than the equivalent expression for the differential operator. For the differential operator we have

$$\frac{d^k(u(x)v(x))}{dx^k} = \sum_{j=0}^{k} \binom{k}{j} u^{(k-j)}(x)v^{(j)}(x) \tag{1.12}$$

The equivalent expression for the difference operator is given below; details of the derivation are given in Appendix A.

$$\Delta^k(u_n v_n) = k! \sum_{n=0}^{k} \sum_{m=0}^{k} \frac{\Delta^n u_n \Delta^m v_n}{(k-n)!(k-m)!(n+m-k)!}$$

in which it is understood that terms vanish when $k > n + m$. Alternatively, we may write

$$\Delta^k(u_n v_n) = k! \sum_{n=0}^{k} \sum_{m=0}^{k} \frac{\Delta^{k-n} u_n \Delta^{k-m} v_n}{n!m!(k-n-m)!}$$

in which terms vanish when $k < n + m$.

Throughout this work we will frequently write sums and products. In this connection we adopt the convention that

$$\prod_{k=0}^{n-1} \equiv 1 \quad \text{for} \quad n = 0 \tag{1.13}$$

and

$$\sum_{k=0}^{n-1} \equiv 0 \quad \text{for} \quad n = 0. \tag{1.14}$$

Chapter 2
Solution of Homogeneous and Inhomogeneous Linear Equations

Although we are concerned primarily with equations (differential and difference) of first and second order, the analysis of these equations applies equally to equations of higher order. The methods for dealing with these equations is in fact best elucidated by considering the nth order equations and then giving the results for the first and second order equations as specific examples. We first present the analysis for differential equations and then follow with the analysis for difference equations.

The idea essential to most of the methods for solving an inhomogeneous equation of nth order is that if one knows a number, m, $(1 \leq m \leq n)$ of linearly independent solutions of the homogeneous equation, then one can derive a linear equation of order $n - m$. For the case in which one knows one solution of the homogeneous equation $(m = 1)$, the method is referred to as "reduction of order". This is particularly useful if one starts with a second order homogeneous or inhomogeneous equation, resulting in a first order equation which can then be solved directly in closed form. For the case in which one knows all n linearly independent solutions of the nth order homogeneous equation, the method (discussed earlier by Lagrange) is referred to as "variation of parameters" or "variation of constants"—a characterization that will become clear shortly. One then obtains the solution to the inhomogeneous equation in terms of the n linearly independent solutions of the homogeneous equation. The case of an intermediate m, $1 < m < n$, has been treated succinctly in a short article by Locke [31], linking the particular cases $m = 1$ (reduction of order) and $m = n$ (variation of parameters). An alternative approach is given in [18]. An analysis for arbitrary m, $1 \leq m \leq n$, in which the nth order equation is transformed into a first order matrix equation, is given in [20].

The nth order linear homogeneous differential equation has the form

$$Ly(x) \equiv a_n(x)y^{(n)}(x) + a_{n-1}(x)y^{(n-1)}(x) + \cdots + a_0(x)y(x)$$

$$= \sum_{i=0}^{n} a_i(x)y^{(i)}(x) = 0 \tag{2.1}$$

© Springer International Publishing Switzerland 2016
L.C. Maximon, *Differential and Difference Equations*,
DOI 10.1007/978-3-319-29736-1_2

and the corresponding inhomogeneous equation is

$$\sum_{i=0}^{n} a_i(x) y^{(i)}(x) = f(x) \tag{2.2}$$

where

$$y^{(i)}(x) \equiv \frac{d^i y(x)}{dx^i}. \tag{2.3}$$

The corresponding Nth order linear homogeneous and inhomogeneous difference equations are

$$Ly(n) \equiv p_N(n)y(n+N) + p_{N-1}(n)y(n+N-1) + \cdots + p_0(n)y(n)$$

$$= \sum_{i=0}^{N} p_i(n)y(n+i) = 0 \tag{2.4}$$

and

$$\sum_{i=0}^{N} p_i(n)y(n+i) = q_n \tag{2.5}$$

respectively. Here the coefficients $p_i(n)$ are often also functions of independent parameters. We note that these two Nth order difference equations can be written in a form similar to that for the differential equations using the difference operator $\Delta y(n) = y(n+1) - y(n)$. From (1.9) we have

$$\sum_{i=0}^{N} p_i(n)y(n+i) = \sum_{i=0}^{N} p_i(n) \sum_{j=0}^{i} \binom{i}{j} \Delta^j y(n)$$

$$= \sum_{j=0}^{N} \Delta^j y(n) \sum_{i=j}^{N} \binom{i}{j} p_i(n). \tag{2.6}$$

We can then write the homogeneous and inhomogeneous difference equations in the form

$$\sum_{j=0}^{N} r_j(n) \Delta^j y(n) = 0 \tag{2.7}$$

and

$$\sum_{j=0}^{N} r_j(n) \Delta^j y(n) = q_n \tag{2.8}$$

respectively, where

$$r_j(n) = \sum_{i=j}^{N} \binom{i}{j} p_i(n).$$ (2.9)

2.1   Variation of Constants

We start with an analysis of the method of variation of constants since it provides the clearest understanding of the essential aspects of the method.

2.1.1   Inhomogeneous Differential Equations

As given above, the nth order homogeneous differential equation is

$$\sum_{j=0}^{n} a_j(x) y^{(j)}(x) = 0$$ (2.10)

and the corresponding inhomogeneous equation is

$$\sum_{j=0}^{n} a_j(x) y^{(j)}(x) = f(x).$$ (2.11)

We assume that the solution to the inhomogeneous equation, $y(x)$, and its $n-1$ derivatives, can be given in terms of the n linearly independent solutions of the homogeneous equation, $u_k(x)$, $(k = 1, 2, \ldots, n)$ by

$$y^{(j)}(x) = \sum_{k=1}^{n} c_k(x) u_k^{(j)}(x) \qquad j = 0, 1, \ldots, n-1.$$ (2.12)

We then have n linear equations for the n functions $c_k(x)$. They do not define the functions $c_k(x)$, but merely relate them to the derivatives $y^{(j)}(x)$, which is possible given the linear independence of the functions $u_k(x)$. We note that if the $c_k(x)$ are constants, then $y(x)$ as defined by this equation is a solution of the *homogeneous* equation. By allowing the $c_k(x)$ to vary (i.e., to be functions of x), we can determine them so that $y(x)$ is a solution of the *inhomogeneous* equation, whence the name variation of constants, or variation of parameters. An equation for the c_k is obtained by substituting (2.12) into (2.2); still required is $y^{(n)}(x)$. Differentiating (2.12) for $j = n-1$ we have

$$y^{(n)}(x) = \sum_{k=1}^{n} c_k(x) u_k^{(n)}(x) + \sum_{k=1}^{n} c_k'(x) u_k^{(n-1)}(x). \tag{2.13}$$

Substituting (2.12) and (2.13) in (2.2) then gives

$$f(x) = \sum_{j=0}^{n-1} a_j(x) \sum_{k=1}^{n} c_k(x) u_k^{(j)}(x) + a_n(x) \left[\sum_{k=1}^{n} c_k(x) u_k^{(n)}(x) + \sum_{k=1}^{n} c_k'(x) u_k^{(n-1)}(x) \right]$$

$$= \sum_{j=0}^{n} a_j(x) \sum_{k=1}^{n} c_k(x) u_k^{(j)}(x) + a_n(x) \sum_{k=1}^{n} c_k'(x) u_k^{(n-1)}(x). \tag{2.14}$$

Interchanging the order of summation in the first term here we have

$$\sum_{j=0}^{n} a_j(x) \sum_{k=1}^{n} c_k(x) u_k^{(j)}(x) = \sum_{k=1}^{n} c_k(x) \sum_{j=0}^{n} a_j(x) u_k^{(j)}(x) = 0 \tag{2.15}$$

since the $u_k(x)$ are solutions of the homogeneous equation (2.1). We now have one equation for the first derivative of the n functions $c_k(x)$:

$$\sum_{k=1}^{n} c_k'(x) u_k^{(n-1)}(x) = \frac{f(x)}{a_n(x)} \equiv g_n(x). \tag{2.16}$$

The remaining $n - 1$ equations defining the functions c_k' follow from (2.12): Differentiation of (2.12) gives

$$y^{(j+1)}(x) = \sum_{k=1}^{n} c_k(x) u_k^{(j+1)}(x) + \sum_{k=1}^{n} c_k'(x) u_k^{(j)}(x). \tag{2.17}$$

Here, from (2.12), for $j = 0, 1, \ldots, n - 2$, the first sum on the right hand side is $y^{(j+1)}(x)$, from which

$$\sum_{k=1}^{n} c_k'(x) u_k^{(j)}(x) = 0, \qquad j = 0, 1, \ldots, n - 2. \tag{2.18}$$

Equations (2.16) and (2.18) now give n equations for the n functions $c_k'(x)$ which can then be integrated to give $c_k(x)$.

All of the results just given can be obtained more succinctly by formulating the matrix equivalent of these equations, giving a first order matrix differential equation. To that end, we define the Wronskian matrix, $\mathbf{W}(x)$,

$$\mathbf{W}(x) = \begin{pmatrix} u_1(x) & u_2(x) & \cdots & u_n(x) \\ u_1^{(1)}(x) & u_2^{(1)}(x) & \cdots & u_n^{(1)}(x) \\ \vdots & \vdots & \vdots & \vdots \\ u_1^{(n-1)}(x) & u_2^{(n-1)}(x) & \cdots & u_n^{(n-1)}(x) \end{pmatrix} \tag{2.19}$$

and the column vectors $\mathbf{c}(x)$, $\mathbf{g}(x)$

$$\mathbf{c}(x) = \begin{pmatrix} c_1(x) \\ c_2(x) \\ \vdots \\ c_n(x) \end{pmatrix} \tag{2.20}$$

$$\mathbf{g}(x) = \begin{pmatrix} 0 \\ 0 \\ \vdots \\ g_n(x) \end{pmatrix} \tag{2.21}$$

and

$$\mathbf{y}(x) = \begin{pmatrix} y_1(x) \\ y_2(x) \\ \vdots \\ y_n(x) \end{pmatrix} \tag{2.22}$$

in which the components $y_j(x)$ are given in terms of the derivatives of the solution to the inhomogeneous equation (2.11) by

$$y_1(x) = y(x), \quad y_2(x) = y^{(1)}(x), \ldots, y_n(x) = y^{(n-1)}(x) \tag{2.23}$$

from which

$$y_j'(x) = y_{j+1}(x), \qquad j = 1, 2, \ldots, n-1 \tag{2.24}$$

The inhomogeneous equation (2.11) may then be written in the form

$$a_n(x)y_n'(x) + \sum_{j=0}^{n-1} a_j(x)y_{j+1}(x) = f(x) \tag{2.25}$$

or

$$y_n'(x) = -\sum_{j=0}^{n-1} b_j y_{j+1}(x) + g_n(x) \tag{2.26}$$

where

$$b_j = b_j(x) = a_j(x)/a_n(x), \qquad g_n(x) = f(x)/a_n(x). \tag{2.27}$$

Taken together, Eqs. (2.24) and (2.26) then give $y'_j(x)$ in terms of $y_j(x)$ for $j = 1, 2, \ldots n$ and can be written in matrix form:

$$\mathbf{y}' = \mathbf{B}\mathbf{y} + \boldsymbol{g} \tag{2.28}$$

where

$$\mathbf{B}(x) = \begin{pmatrix} 0 & 1 & \cdots & 0 \\ \vdots & \vdots & \vdots & \vdots \\ 0 & 0 & \cdots & 1 \\ -b_0 & -b_1 & \cdots & -b_{n-1} \end{pmatrix} \tag{2.29}$$

Equation (2.12), defining the $n - 1$ derivatives of $y(x)$, then takes the simple form

$$\mathbf{y} = \mathbf{W}\mathbf{c} \tag{2.30}$$

when written in terms of the Wronskian matrix $\mathbf{W}(x)$. Substituting this in the matrix form of the inhomogeneous equation, (2.28), then gives

$$(\mathbf{W}\mathbf{c})' = \mathbf{W}'\mathbf{c} + \mathbf{W}\mathbf{c}' = \mathbf{B}\mathbf{W}\mathbf{c} + \boldsymbol{g} \tag{2.31}$$

From the homogeneous equation satisfied by $u_k(x)$:

$$u_k^{(n)} = -\sum_{j=0}^{n-1} b_j u_k^{(j)}, \quad k = 1, 2, \ldots, n \tag{2.32}$$

we have

$$\mathbf{B}(x)\mathbf{W}(x) = \mathbf{W}'(x), \tag{2.33}$$

which, when substituted in (2.31), gives

$$\mathbf{W}\mathbf{c}' = \boldsymbol{g}, \tag{2.34}$$

which is equivalent to Eqs. (2.16) and (2.18). Integration of this equation gives

$$\mathbf{c}(x) = \int^x \mathbf{W}^{-1}(x')g(x')dx', \tag{2.35}$$

and from (2.30),

$$\mathbf{y}(x) = \mathbf{W}(x) \int^x \mathbf{W}^{-1}(x')g(x')dx' \tag{2.36}$$

If we have an initial value problem in which $y(x)$ and its n derivatives are specified at a point $x = x_0$, then integration of equation (2.34) gives

$$\mathbf{c}(x) = \mathbf{c}(x_0) + \int_{x_0}^{x} \mathbf{W}^{-1}(x')g(x')dx' \tag{2.37}$$

Multiplying both sides of this equation by $\mathbf{W}(x)$ and using $\mathbf{c}(x_0) = \mathbf{W}^{-1}(x_0)\mathbf{y}(x_0)$ from (2.30) we have the solution to the inhomogeneous equation (2.11) in matrix form:

$$\mathbf{y}(x) = \mathbf{W}(x)\left(\mathbf{W}^{-1}(x_0)\mathbf{y}(x_0) + \int_{x_0}^{x} \mathbf{W}^{-1}(x')g(x')dx'\right). \tag{2.38}$$

We note that the three essential equations in this analysis are (2.28), $\mathbf{y}' = \mathbf{B}\mathbf{y} + g$, which defines the inhomogeneous equation and is equivalent to Eq. (2.2); (2.30), $\mathbf{y} = \mathbf{W}\mathbf{c}$, which relates the functions $c_k(x)$ to the derivatives of $y(x)$ and is equivalent to Eq. (2.12); and (2.33), $\mathbf{B}\mathbf{W} = \mathbf{W}'$, which gives the homogeneous equation satisfied by its solutions $u_k(x)$ and is equivalent to Eq. (2.32). With only minor notational modifications, these three equations form the basis of the analysis for difference equations (note Eqs. (2.64), (2.66) and (2.69)).

An alternate but completely equivalent approach to the solution of the nth order linear inhomogeneous equation, (2.28),

$$\mathbf{y}' = \mathbf{B}\mathbf{y} + g, \tag{2.39}$$

is provided by consideration of the Wronskian (in the case of the differential equation) and the Casoratian (in the case of the difference equation). We start from Eq. (2.34),

$$\mathbf{W}\mathbf{c}' = g, \tag{2.40}$$

The solution to this set of equations is given by Cramer's rule (see Appendix E), by which the column vector $\mathbf{g}(x)$, (2.21), replaces the jth column in the Wronskian matrix $\mathbf{W}(x)$, (2.19). The elements $c_j'(x)$ are then given by

$$c_j'(x) = \frac{1}{W(x)} \begin{vmatrix} u_1 & \cdots & u_{j-1} & 0 & u_{j+1} & \cdots u_n \\ u_1^{(1)} & \cdots & u_{j-1}^{(1)} & 0 & u_{j+1}^{(1)} & \cdots u_n^{(1)} \\ \vdots & \vdots & \vdots & \vdots & \vdots & \vdots & \vdots \\ u_1^{(n-2)} & \cdots & u_{j-1}^{(n-2)} & 0 & u_{j+1}^{(n-2)} & \cdots u_n^{(n-2)} \\ u_1^{(n-1)} & \cdots & u_{j-1}^{(n-1)} & g_n & u_{j+1}^{(n-1)} & \cdots u_n^{(n-1)} \end{vmatrix} \tag{2.41}$$

where $u_k = u_k(x)$, $(k = 1, 2, \ldots, n)$, are the n linearly independent solutions of (2.1) and W is the determinant of the Wronskian matrix (2.19). (See Appendix B for $j = 1$ and $j = n$.) Expanding the determinant (2.41) in the elements of the jth column, $c_j'(x)$ can be expressed in terms of an $(n-1) \times (n-1)$ determinant:

$$c'_j(x) = (-1)^{n+j} \frac{g_n(x)}{W(x)} \begin{vmatrix} u_1 & \cdots & u_{j-1} & u_{j+1} & \cdots & u_n \\ u_1^{(1)} & \cdots & u_{j-1}^{(1)} & u_{j+1}^{(1)} & \cdots & u_n^{(1)} \\ \vdots & & \vdots & \vdots & & \vdots \\ u_1^{(n-2)} & \cdots & u_{j-1}^{(n-2)} & u_{j+1}^{(n-2)} & \cdots & u_n^{(n-2)} \end{vmatrix} \qquad (2.42)$$

(See Appendix B for $j = 1$ and $j = n$.)

The matrix solution to the inhomogeneous equation may then be written if one constructs a column vector $\mathbf{c}'(x)$ whose elements are the $c'_j(x)$ for $j = 1, 2, \ldots, n$. Equation (2.30), $\mathbf{y}(x) = \mathbf{W}(x)\mathbf{c}(x)$, then gives

$$\mathbf{y}(x) = \mathbf{W}(x) \int_{x_0}^{x} \mathbf{c}'(x')dx' \qquad (2.43)$$

The first element in the column vector $\mathbf{y}(x)$ gives the function $y(x)$:

$$y(x) = \sum_{j=1}^{n} u_j(x) \int_{x_0}^{x} c'_j(x')dx' \qquad (2.44)$$

with $c'_j(x)$ given in (2.42). This provides a particular solution, to which an arbitrary solution to the homogeneous equation may be added to satisfy boundary conditions.

2.1.2   Inhomogeneous Difference Equations

We now look at the equivalent analysis for the Nth order inhomogeneous difference equation

$$p_N(n)y(N+n) + p_{N-1}(n)y(N+n-1) + \cdots + p_0(n)y(n) = \sum_{j=0}^{N} p_j(n)y(n+j) = q_N(n) \qquad (2.45)$$

The Nth order homogeneous equation is

$$\sum_{j=0}^{N} p_j(n)y(n+j) = 0, \qquad (2.46)$$

for which the N linearly independent solutions are denoted by $u_k(n)$, $(k = 1, 2, \ldots, N)$, that is,

$$\sum_{j=0}^{N} p_j(n)u_k(n+j) = 0, \quad k = 1, 2, \ldots, N \qquad (2.47)$$

As with the differential equation, we assume that the solution to the inhomogeneous equation, $y(n)$, and the succeeding $N - 1$ terms $y(n + 1)$, $y(n + 2), \ldots, y(n + N - 1)$ can be given in terms of the N linearly independent solutions $u_k(n)$ of the homogeneous equation by

$$y(n + j) = \sum_{k=1}^{N} c_k(n)u_k(n + j) \qquad j = 0, 1, \ldots, N - 1 \qquad (2.48)$$

We thus have N linear equations determining the N functions, $c_k(n)$, which is possible given the linear independence of the functions $u_k(n)$. We note that if the c_k are constants, then $y(n)$ as defined by this equation is a solution of the *homogeneous* equation. By allowing the c_k to vary (i.e., to be functions of n), we can determine them so that $y(n)$ is a solution of the *inhomogeneous* equation. From (2.48) we can thus write

$$y(n + j + 1) = \sum_{k=1}^{N} c_k(n)u_k(n + j + 1) \quad \text{for} \quad j = 0, 1, \ldots, N - 2 \qquad (2.49)$$

as well as

$$y(n + j + 1) = \sum_{k=1}^{N} c_k(n + 1)u_k(n + j + 1) \quad \text{for} \quad j = 0, 1, \ldots, N - 2 \qquad (2.50)$$

from which we have the $N - 1$ equations

$$\sum_{k=1}^{N} \Delta c_k(n)u_k(n + j + 1) = 0, \qquad j = 0, 1, \ldots, N - 2 \qquad (2.51)$$

From (2.48) for $j = N - 1$ we have

$$y(n + N - 1) = \sum_{k=1}^{N} c_k(n)u_k(n + N - 1) \qquad (2.52)$$

from which

$$y(n + N) = \sum_{k=1}^{N} c_k(n + 1)u_k(n + N)$$

$$= \sum_{k=1}^{N} c_k(n)u_k(n + N) + \sum_{k=1}^{N} \Delta c_k(n)u_k(n + N) \qquad (2.53)$$

Substituting (2.48) and (2.53) in (2.45) we then have

$$\sum_{j=0}^{N} p_j(n) \sum_{k=1}^{N} c_k(n) u_k(n+j) + p_N(n) \sum_{k=1}^{N} \Delta c_k(n) u_k(n+N) = q_N(n) \quad (2.54)$$

Inverting the order of summation in the first term here, we see that this term vanishes since the $u_k(n)$ satisfy the homogeneous equation (2.46). We thus have

$$\sum_{k=1}^{N} \Delta c_k(n) u_k(n+N) = \frac{q_N(n)}{p_N(n)} \equiv h_N(n) \quad (2.55)$$

Equation (2.55) together with (2.51) for $j = 0, 1, \ldots, N-2$ then give N equations for the N differences $\Delta c_k(n)$ which can then be summed to give the functions $c_k(n)$.

We can now formulate the entire analysis for difference equations in terms of matrices, in a manner quite similar to that for differential equations, giving a first order matrix difference equation. To that end we define the Casoratian matrix:

$$\mathbf{K}(n) = \begin{pmatrix} u_1(n) & u_2(n) & \cdots u_N(n) \\ u_1(n+1) & u_2(n+1) & \cdots u_N(n+1) \\ \vdots & \vdots & \vdots \ \vdots \\ u_1(n+N-1) & u_2(n+N-1) & \cdots u_N(n+N-1) \end{pmatrix} \quad (2.56)$$

and the column vectors $\mathbf{c}(n)$, $\mathbf{h}(n)$

$$\mathbf{c}(n) = \begin{pmatrix} c_1(n) \\ c_2(n) \\ \vdots \\ c_N(n) \end{pmatrix} \quad (2.57)$$

$$\mathbf{h}(n) = \begin{pmatrix} 0 \\ 0 \\ \vdots \\ h_N(n) \end{pmatrix} \quad (2.58)$$

and

$$\mathbf{y}(n) = \begin{pmatrix} y_1(n) \\ y_2(n) \\ \vdots \\ y_N(n) \end{pmatrix} \quad (2.59)$$

in which the components $y_j(n)$ are given in terms of the solution of the inhomogeneous equation (2.45) for successive indices by

$$y_j(n) = y(j+n-1), \qquad j = 1, 2, \ldots, N \quad (2.60)$$

We can then write the inhomogeneous equation (2.45) in the form

$$y(N + n) + b_{N-1}y(N + n - 1) + \cdots + b_0 y(n) = h_N(n) \qquad (2.61)$$

or

$$y(N + n) = -\sum_{j=0}^{N-1} b_j y_{j+1}(n) + h_N(n) \qquad (2.62)$$

where

$$b_j = b_j(n) = p_j(n)/p_N(n), \qquad h_N(n) = q_N(n)/p_N(n) \qquad (2.63)$$

This may then be written in matrix form (cf. Eq. (2.28) for differential equations) as

$$\mathbf{y}(n + 1) = \mathbf{B}(n)\mathbf{y}(n) + \mathbf{h}(n) \qquad (2.64)$$

where

$$\mathbf{B}(n) = \begin{pmatrix} 0 & 1 & \cdots & 0 \\ \vdots & \vdots & \vdots & \vdots \\ 0 & 0 & \cdots & 1 \\ -b_0 & -b_1 & \cdots & -b_{N-1} \end{pmatrix} \qquad (2.65)$$

Equation (2.48) giving the N terms $y(n + j)$ for $j = 0, 1, \ldots, N - 1$ then takes the simple matrix form (cf. Eq. (2.30) for differential equations)

$$\mathbf{y}(n) = \mathbf{K}(n)\mathbf{c}(n) \qquad (2.66)$$

Substituting (2.66) in (2.64) we then have

$$\mathbf{y}(n + 1) = \mathbf{B}(n)\mathbf{K}(n)\mathbf{c}(n) + \mathbf{h}(n) \qquad (2.67)$$

Here

$$\mathbf{B}(n)\mathbf{K}(n) = \begin{pmatrix} 0 & 1 & \cdots & 0 \\ \vdots & \vdots & \vdots & \vdots \\ 0 & 0 & \cdots & 1 \\ -b_0 & -b_1 & \cdots & -b_{n-1} \end{pmatrix} \begin{pmatrix} u_1(n) & u_2(n) & \cdots u_N(n) \\ u_1(n + 1) & u_2(n + 1) & \cdots u_N(n + 1) \\ \vdots & \vdots & \vdots \vdots \\ u_1(n + N - 1) & u_2(n + N - 1) & \cdots u_N(n + N - 1) \end{pmatrix}$$

$$= \begin{pmatrix} u_1(n + 1) & u_2(n + 1) & \cdots u_N(n + 1) \\ u_1(n + 2) & u_2(n + 2) & \cdots u_N(n + 2) \\ \vdots & \vdots & \vdots \vdots \\ u_1(n + N) & u_2(n + N) & \cdots u_N(n + N) \end{pmatrix} \qquad (2.68)$$

Thus

$$\mathbf{B}(n)\mathbf{K}(n) = \mathbf{K}(n+1) \tag{2.69}$$

(cf. Eq. (2.33) for differential equations.) The last line in $\mathbf{K}(n+1)$ follows from the homogeneous equation (2.47) satisfied by the functions $u_k(n)$:

$$u_k(N+n) = -b_0(n)u_k(n) - b_1(n)u_k(n+1) - \cdots - b_N(n)u_k(n+N-1) \tag{2.70}$$

We note that Eqs. (2.64), (2.66) and (2.69) for difference equations correspond respectively to Eqs. (2.28), (2.30) and (2.33) for differential equations. Substituting (2.69) in (2.64) then gives

$$\mathbf{y}(n+1) = \mathbf{K}(n+1)\mathbf{c}(n) + \mathbf{h}(n) \tag{2.71}$$

However, from (2.66) we can also write

$$\mathbf{y}(n+1) = \mathbf{K}(n+1)\mathbf{c}(n+1) \tag{2.72}$$

so that from the last two equations we have

$$\mathbf{K}(n+1)\Delta\mathbf{c}(n) = \mathbf{h}(n), \tag{2.73}$$

from which

$$\Delta\mathbf{c}(n) = \mathbf{K}^{-1}(n+1)\mathbf{h}(n) \tag{2.74}$$

and

$$\mathbf{c}(n+1) = \mathbf{c}(0) + \sum_{j=0}^{n} \mathbf{K}^{-1}(j+1)\mathbf{h}(j) \tag{2.75}$$

(cf. (2.37) for differential equations.) Here, the term $\mathbf{c}(0)$ adds an arbitrary solution of the homogeneous equation and is determined by the initial conditions. Equation (2.73) is the matrix form of Eq. (2.55) together with (2.51) for $j = 0, 1, \ldots, N-2$. Equations (2.73), (2.74) and (2.75) given above for difference equations correspond to Eqs. (2.34) and (2.37) for differential equations. Writing (2.75) with $n+1$ replaced by n we have

$$\mathbf{c}(n) = \mathbf{c}(0) + \sum_{j=1}^{n} \mathbf{K}^{-1}(j)\mathbf{h}(j-1) \tag{2.76}$$

Multiplying both sides of this equation by $\mathbf{K}(n)$ and using (2.66) (from which $\mathbf{c}(0) = \mathbf{K}^{-1}(0)\mathbf{y}(0)$) we have the solution to the inhomogeneous equation (2.64) in matrix form:

$$\mathbf{y}(n) = \mathbf{K}(n)\left(\mathbf{K}^{-1}(0)\mathbf{y}(0) + \sum_{j=1}^{n} \mathbf{K}^{-1}(j)\mathbf{h}(j-1)\right) \tag{2.77}$$

Similar to the case for differential equations, we note that the three essential equations in the analysis for difference equations are (2.64), $\mathbf{y}(n+1) = \mathbf{B}(n)\mathbf{y}(n) + \mathbf{h}(n)$, which defines the inhomogeneous equation and is equivalent to Eq. (2.45); (2.66), $\mathbf{y}(n) = \mathbf{K}(n)\mathbf{c}(n)$, which relates the N functions $c_k(n)$ to N successive terms $\mathbf{y}(n+j)$ for $j = 0, 1, \ldots, N-1$ and is equivalent to Eq. (2.48); and (2.69), $\mathbf{B}(n)\mathbf{K}(n) = \mathbf{K}(n+1)$, which gives the homogeneous equation satisfied by its solutions $u_k(n)$ and is equivalent to Eq. (2.47).

An alternate but completely equivalent approach to the solution of the Nth order linear inhomogeneous equation, (2.64),

$$\mathbf{y}(n+1) = \mathbf{B}(n)\mathbf{y}(n) + \mathbf{h}(n) \tag{2.78}$$

is provided by consideration of the Casoratian in the case of the difference equation. We start from Eq. (2.73),

$$\mathbf{K}(n+1)\Delta\mathbf{c}(n) = \mathbf{h}(n), \tag{2.79}$$

Replacing $n+1$ by n we have

$$\mathbf{K}(n)\Delta\mathbf{c}(n-1) = \mathbf{h}(n-1), \tag{2.80}$$

The solution to this matrix equation is given by Cramer's rule, from which the elements $\Delta c_j(n-1)$ of the column vector $\Delta\mathbf{c}(n-1)$ for $j = 1, 2, \ldots, N$ are given by

$$\Delta c_j(n-1) = \frac{1}{K(n)} \begin{vmatrix} u_1(n) & \cdots u_{j-1}(n) & 0 & u_{j+1}(n) & \cdots u_N(n) \\ u_1(n+1) & \cdots u_{j-1}(n+1) & 0 & u_{j+1}(n+1) & \ddots, \ u_N(n+1) \\ \vdots & \vdots & \vdots & \vdots & \vdots & \vdots \\ u_1(n+N-2) & \cdots u_{j-1}(n+N-2) & 0 & u_{j+1}(n+N-2) & \cdots u_N(n+N-2) \\ u_1(n+N-1) & \cdots u_{j-1}(n+N-1) & h_N(n-1) & u_{j+1}(n+N-1) & \cdots u_N(n+N-1), \end{vmatrix} \tag{2.81}$$

where $u_j = u_j(n)$, $j = 1, 2, \ldots, N$, are the N linearly independent solutions of (2.4) and $K(n)$ is the determinant of the Casoratian matrix (2.56). Expanding the determinant (2.81) in the elements of the jth column, the elements $\Delta c_j(n-1)$ can be expressed in terms of an $(N-1) \times (N-1)$ determinant:

$$\Delta c_j(n-1) = (-1)^{N+j}\frac{h_N(n-1)}{K(n)} \begin{vmatrix} u_1(n) & \cdots u_{j-1}(n) & u_{j+1}(n) & \cdots u_N(n) \\ u_1(n+1) & \cdots u_{j-1}(n+1) & u_{j+1}(n+1) & \cdots u_N(n+1) \\ \vdots & \vdots & \vdots & \vdots & \vdots & \vdots \\ u_1(n+N-2) & \cdots u_{j-1}(n+N-2) & u_{j+1}(n+N-2) & \cdots u_N(n+N-2) \end{vmatrix} \tag{2.82}$$

It is clear that the determinant as written in (2.82) is valid for $j = 2, 3, \ldots, N-1$. For $j = 1$ and $j = N$ (as well as for all $1 \le j \le N$) one must simply omit the jth column. The matrix solution to the inhomogeneous equation may then be written from the column vector $\Delta \mathbf{c}(n-1)$ whose elements are the $\Delta c_j(n-1)$ given in (2.82) for $j = 1, 2, \ldots, N$. Equation (2.66), $\mathbf{y}(n) = \mathbf{K}(n)\mathbf{c}(n)$, then gives $\mathbf{y}(n)$, with

$$\mathbf{y}(n) = \mathbf{K}(n)\left(\mathbf{K}^{-1}(0)y(0) + \sum_{n'=1}^{n} \Delta \mathbf{c}(n'-1) \right) \tag{2.83}$$

The first term in parentheses gives a solution of the homogeneous equation. Therefore a particular matrix solution of the inhomogeneous equation is given by

$$\mathbf{y}(n) = \mathbf{K}(n) \sum_{n'=1}^{n} \Delta \mathbf{c}(n'-1) \tag{2.84}$$

The first element of this matrix equation gives the function $y(n)$:

$$y(n) = \sum_{j=1}^{N} u_j(n) \sum_{n'=1}^{n} \Delta c_j(n'-1) \tag{2.85}$$

with $\Delta c_j(n'-1)$ given by (2.82). This provides a particular solution to which an arbitrary solution to the homogeneous equation may be added to satisfy boundary conditions.

2.2 Reduction of the Order When One Solution to the Homogeneous Equation Is Known

The present method reduces the order of an nth order linear operator, giving an operator of order $n-1$ when one solution to the homogeneous equation is known. Thus, an nth order homogeneous equation $Ly = 0$ is transformed into a homogeneous equation $\mathscr{L}w = 0$ of order $n-1$ in w; an nth order inhomogeneous equation $Ly = f$ is transformed into an inhomogeneous equation $\mathscr{L}w = f$ of order $n-1$ in w. (In particular, for a second order equation we obtain a first order equation, which is then soluble in closed form.) The details in the analysis of differential and difference equations are quite similar, and the approach is the same as that given earlier in connection with the method of variation of constants (cf. (2.12)): By writing the dependent variable ($y(x)$ or $y(n)$) as the product of two functions,

$$y(x) = c(x)u(x) \tag{2.86}$$

or

$$y(n) = c(n)u(n), \tag{2.87}$$

one of which satisfies the homogeneous equation ($Lu(x) = 0$ or $Lu(n) = 0$, respectively), one can write the original equation in a form such that only derivatives (or differences) of the unknown function ($c(x)$ or $c(n)$) appear. Then, defining

$$w(x) = c'(x) \tag{2.88}$$

and

$$w(n) = \Delta c(n) = c(n + 1) - c(n), \tag{2.89}$$

the order of the equation for $w(x)$ or $w(n)$ is less by one than that of the original equation.

We start by considering the nth order differential operator given in (2.1), viz.,

$$Ly(x) \equiv a_n(x)y^{(n)}(x) + a_{n-1}(x)y^{(n-1)}(x) + \cdots + a_0(x)y(x) = \sum_{j=0}^{n} a_j(x)y^{(j)}(x) \tag{2.90}$$

Writing

$$y(x) = c(x)u(x) \tag{2.91}$$

where $u(x)$ is assumed to be a known solution of

$$Lu(x) = \sum_{j=0}^{n} a_j(x)u^{(j)}(x) = 0, \tag{2.92}$$

we have

$$
\begin{aligned}
y^{(k)}(x) &= \frac{d^k(c(x)u(x))}{dx^k} \\
&= \sum_{j=0}^{k} \binom{k}{j} c^{(j)}(x)u^{(k-j)}(x)
\end{aligned}
\tag{2.93}
$$

and from (2.1),

$$
\begin{aligned}
Ly(x) &= \sum_{k=0}^{n} a_k(x)y^{(k)}(x) \\
&= \sum_{k=0}^{n} a_k(x) \sum_{j=0}^{k} \binom{k}{j} c^{(j)}(x)u^{(k-j)}(x) \\
&= \sum_{j=0}^{n} c^{(j)}(x) \sum_{k=j}^{n} \binom{k}{j} a_k(x)u^{(k-j)}(x) \\
&= \sum_{j=1}^{n} c^{(j)}(x) \sum_{k=j}^{n} \binom{k}{j} a_k(x)u^{(k-j)}(x) + c(x) \sum_{k=0}^{n} a_k(x)u^{(k)}(x)
\end{aligned}
\tag{2.94}
$$

The last sum here is zero from (2.92). Then, defining

$$w(x) \equiv c^{(1)}(x), \tag{2.95}$$

we obtain a differential operator of order $n - 1$ in $w(x)$:

$$\sum_{j=1}^{n} c^{(j)}(x) \sum_{k=j}^{n} \binom{k}{j} a_k(x) u^{(k-j)}(x) = \sum_{j=1}^{n} w^{(j-1)}(x) \sum_{k=j}^{n} \binom{k}{j} a_k(x) u^{(k-j)}(x)$$

$$= \sum_{j=0}^{n-1} w^{(j)}(x) \sum_{k=j+1}^{n} \binom{k}{j+1} a_k(x) u^{(k-j-1)}(x)$$

$$= \sum_{j=0}^{n-1} w^{(j)}(x) \sum_{k=j}^{n-1} \binom{k+1}{j+1} a_{k+1}(x) u^{(k-j)}(x)$$

$$= \mathscr{L}w \tag{2.96}$$

We next look at the analogous procedure for an Nth order linear homogeneous difference operator, given in (2.4), viz.,

$$Ly(n) \equiv p_N(n)y(n + N) + p_{N-1}(n)y(n + N - 1) + \cdots + p_0(n)y(n) \tag{2.97}$$

Again, we write the solution of this equation as the product of two functions:

$$y(n) = c(n)u(n) \tag{2.98}$$

where we assume $u(n)$ to be a known solution of the homogeneous equation

$$Lu(n) = p_N(n)u(n + N) + p_{N-1}(n)u(n + N - 1) + \cdots + p_0(n)u(n)$$

$$= \sum_{k=0}^{N} p_k(n)u(n + k)$$

$$= 0 \tag{2.99}$$

The operator (2.97) is then

$$Ly(n) = p_N(n)c(n + N)u(n + N) + p_{N-1}(n)c(n + N - 1)u(n + N - 1)$$

$$+ \cdots + p_0(n)c(n)u(n) \tag{2.100}$$

Applying (1.9) to the function $c(n+k)$, this operator can be written in the form

$$\sum_{k=0}^{N} p_k(n)u(n+k) \sum_{j=0}^{k} \binom{k}{j} \Delta^j c(n)$$

$$= \sum_{k=1}^{N} p_k(n)u(n+k) \sum_{j=1}^{k} \binom{k}{j} \Delta^j c(n) + c(n) \sum_{k=0}^{N} p_k(n)u(n+k) \qquad (2.101)$$

As with the differential equation, the last sum in the above equation is zero in view of (2.99), giving

$$Ly(n) = \sum_{k=1}^{N} p_k(n)u(n+k) \sum_{j=1}^{k} \binom{k}{j} \Delta^j c(n) \qquad (2.102)$$

Then, in analogy with (2.95), we define

$$w(n) = \Delta c(n), \qquad (2.103)$$

giving

$$\sum_{k=1}^{N} p_k(n)u(n+k) \sum_{j=1}^{k} \binom{k}{j} \Delta^{j-1} w(n) = \sum_{j=1}^{N} \Delta^{j-1} w(n) \sum_{k=j}^{N} \binom{k}{j} p_k(n)u(n+k)$$

$$= \sum_{j=0}^{N-1} \Delta^j w(n) \sum_{k=j+1}^{N} \binom{k}{j+1} p_k(n)u(n+k)$$

$$= \sum_{j=0}^{N-1} \Delta^j w(n) \sum_{k=j}^{N-1} \binom{k+1}{j+1} p_{k+1}(n)u(n+k+1)$$

$$= \mathscr{L}w(n) \qquad (2.104)$$

which is a difference operator of order $N-1$ in $w(n)$.

2.2.1 Solution of Nth Order Inhomogeneous Equations When m Linearly Independent Solutions of the Homogeneous Equation are Known, Where $1 < m < N$

The two methods—reduction of order and variation of parameters—have been presented separately, since that is how they are generally found in the literature. However, as has been shown in a succinct article by Phil Locke [31], each of these procedures can be viewed as particular limiting cases in the solution of an nth order linear non-homogeneous equation when $m \leq n$ linearly independent solutions of the nth order homogeneous equation are known: $m = 1$ corresponds to reduction of order, $m = n$ corresponds to variation of parameters. Related treatments may be found in [18, Chap. IX, Sect. 3, pp. 319–322] and in [20, Chap. IV, Sect. 3, pp. 49–54].

Chapter 3
First Order Homogeneous and Inhomogeneous Linear Equations

The solution of first order inhomogeneous linear equations provides the simplest example of the method of variation of constants given previously in Chap. 2, Sect. 2.1. We start by considering the first order differential equation. With $n = 1$, the homogeneous and inhomogeneous equations are given by (2.10) and (2.11) respectively:

$$a_1(x)u'(x) + a_0(x)u(x) = 0 \tag{3.1}$$

and

$$a_1(x)y'(x) + a_0(x)y(x) = f(x) \tag{3.2}$$

For the first order equation, the solution to the homogeneous equation follows directly by integration:

$$\frac{u'(x)}{u(x)} = -\frac{a_0(x)}{a_1(x)} \tag{3.3}$$

giving

$$u(x) = A \exp\left(-\int^x \frac{a_0(x')}{a_1(x')} dx'\right) \tag{3.4}$$

Here the lower limit of the integration may be chosen arbitrarily, it merely introduces a constant factor to the solution $u(x)$. The method of variation of constants then gives the solution to the inhomogeneous equation. From (2.12) with $n = 1$ we have (omitting the subscripts on $c(x)$ and $u(x)$)

$$y(x) = c(x)u(x) \tag{3.5}$$

© Springer International Publishing Switzerland 2016
L.C. Maximon, *Differential and Difference Equations*,
DOI 10.1007/978-3-319-29736-1_3

and from (2.16)

$$c'(x)u(x) = \frac{f(x)}{a_1(x)} \tag{3.6}$$

from which

$$c(x) = \int^x \frac{f(x')}{a_1(x')u(x')} \, dx' \tag{3.7}$$

A particular solution to the inhomogeneous equation is then

$$y(x) = u(x) \int^x \frac{f(x')}{a_1(x')u(x')} \, dx' \tag{3.8}$$

A solution determined by its value at a given point $x = x_0$ may be obtained by adding to (3.8) a solution of the homogeneous equation. This may be written succinctly by choosing the lower limit x_0 on the integrals in (3.4) and (3.8), and setting $A = 1$ in (3.4) so that $u(x_0) = 1$. We then have

$$y(x) = u(x) \left[y(x_0) + \int_{x_0}^x \frac{f(x')}{a_1(x')u(x')} \, dx' \right] \tag{3.9}$$

The solutions to the first order homogeneous and inhomogeneous difference equations follow analogously. Referring to (2.46) with $N = 1$, the first order homogeneous equation is

$$p_1(n)u(n+1) + p_0(n)u(n) = 0 \tag{3.10}$$

Writing this in the form

$$u(n+1) = -\frac{p_0(n)}{p_1(n)}u(n) \tag{3.11}$$

the solution follows by iteration, giving

$$u(n) = A(-1)^n \prod_k^{n-1} \frac{p_0(k)}{p_1(k)} \tag{3.12}$$

Here A may be an arbitrary constant or a periodic function of period one. The lower limit of the product may be chosen arbitrarily, merely introducing a constant factor to the solution $u(n)$.

The solution to the inhomogeneous equation

$$p_1(n)y(n+1) + p_0(n)y(n) = q_1(n) \tag{3.13}$$

is again given by the method of variation of constants. From (2.48) we have, with $N = 1$ (omitting subscripts on $c(n)$ and $u(n)$),

$$y(n) = c(n)u(n) \tag{3.14}$$

and from (2.55)

$$\Delta c(n)u(n+1) = \frac{q_1(n)}{p_1(n)} \tag{3.15}$$

from which

$$c(n+1) - c(n) = \frac{q_1(n)}{p_1(n)u(n+1)} \tag{3.16}$$

Summation then gives

$$c(n) = \sum_{k}^{n-1} \frac{q_1(k)}{p_1(k)u(k+1)} \tag{3.17}$$

or, alternatively,

$$c(n) = \sum_{k}^{n-1} \frac{q_1(k)}{p_0(k)u(k)} \tag{3.18}$$

A particular solution to the homogeneous equation is then

$$y(n) = u(n) \sum_{k}^{n-1} \frac{q_1(k)}{p_1(k)u(k+1)} \tag{3.19}$$

A solution determined by its value at a given point $n = n_0$ may be obtained by adding to (3.19) a solution of the homogeneous equation. This may be written succinctly by choosing the lower limit n_0 on the summation (3.19) and the product in (3.12), and setting $A = (-1)^{-n_0}$ in (3.12) so that

$$u(n) = (-1)^{n-n_0} \prod_{k=n_0}^{n-1} \frac{p_0(k)}{p_1(k)} \tag{3.20}$$

and hence

$$u(n_0) = 1 \tag{3.21}$$

We then have

$$y(n) = u(n)\left[y(n_0) + \sum_{k=n_0}^{n-1} \frac{q_1(k)}{p_1(k)u(k+1)}\right] \tag{3.22}$$

We note that the solutions to the first order homogeneous differential and difference equations are given respectively by the Wronskian and Casoratian determinants, given in Appendices C and D. From (C.2) with $n = 1$, the Wronskian satisfies the first order Eq. (3.1), and from (C.1) and (C.3)

$$W(x) = u_1(x) = e^{-\int^x \frac{a_0(x')}{a_1(x')} dx'}, \tag{3.23}$$

given in (3.4).

From (D.2) with $N = 1$, the Casoratian satisfies the first order Eq. (3.10), and from (D.1) and (D.3)

$$\mathcal{C}(n) = u_1(n) = (-1)^n \prod_{k}^{n-1} \frac{p_0(k)}{p_1(k)}, \tag{3.24}$$

given in (3.12).

Chapter 4
Second Order Homogeneous and Inhomogeneous Equations

Second order equations provide an interesting example for comparing the methods of variation of constants and reduction of order. As noted in Chap. 2, Sect. 2, if one solution of the homogeneous equation is known then the method of reduction of order transforms a second order equation into a first order equation, which can then be solved in closed form. If the original second order equation is homogeneous then the transformed first order equation is also homogeneous, and its solution provides a second, linearly independent solution to the second order equation. We first consider the second order homogeneous differential equation

$$Ly(x) = a_2(x)y''(x) + a_1(x)y'(x) + a_0(x)y(x) = 0, \qquad (4.1)$$

and assume that one solution, $u_1(x)$, to the homogeneous equation is known:

$$Lu_1(x) = 0. \qquad (4.2)$$

Then from (2.91), by writing

$$y(x) = c(x)u_1(x), \qquad (4.3)$$

the second degree operator $Ly(x)$ is transformed into a first degree operator $\mathscr{L}w$:

$$Ly = \mathscr{L}w \qquad (4.4)$$

where

$$w(x) = c'(x) \qquad (4.5)$$

We denote by $w_0(x)$ the solution of $\mathscr{L}w = 0$, which then determines a second solution to the homogeneous equation $Ly = 0$. From (2.96) the function $w_0(x)$ satisfies the

© Springer International Publishing Switzerland 2016
L.C. Maximon, *Differential and Difference Equations*,
DOI 10.1007/978-3-319-29736-1_4

first order homogeneous equation

$$\mathcal{L}(w_0) = w'_0(x)a_2(x)u_1(x) + w_0(x)\left(a_1(x)u_1(x) + 2a_2(x)u_1'(x)\right) = 0. \quad (4.6)$$

Writing this in the form

$$\frac{w'_0(x)}{w_0(x)} = -\frac{a_1(x)}{a_2(x)} - 2\frac{u_1'(x)}{u_1(x)} \quad (4.7)$$

and integrating, we have

$$w_0(x) = C\frac{\exp\left(-\int^x \frac{a_1(x')}{a_2(x')}\,dx'\right)}{u_1^2(x)} \quad (4.8)$$

which determines a second, linearly independent, solution to the second order homogeneous equation:

$$u_2(x) = u_1(x)\int^x w_0(x')\,dx'. \quad (4.9)$$

That $u_1(x)$ and $c(x)u_1(x)$ are linearly independent follows from the Wronskian determinant

$$W(x) = \begin{vmatrix} u_1(x) & c(x)u_1(x) \\ u_1'(x) & (c(x)u_1(x))' \end{vmatrix} = c'(x)u_1^2(x) = w_0(x)u_1^2(x)$$

$$= C\exp\left(-\int^x \frac{a_1(x')}{a_2(x')}\,dx'\right) > 0 \quad (4.10)$$

The most general solution to the homogeneous equation $Ly = 0$ is then, from (4.2) and (4.9),

$$y(x) = Au_1(x) + Bu_1(x)\int^x w_0(x')\,dx' \quad (4.11)$$

A solution satisfying given initial conditions $y(x_0)$ and $y'(x_0)$ then determines the constants A and B when the lower limit of the integral is chosen to be x_0:

$$Au_1(x_0) + Bu_1(x_0)\int_{x_0}^x w_0(x')\,dx' = y(x_0)$$

$$Au_1'(x_0) + Bu_1(x_0)w_0(x_0) = y'(x_0) \quad (4.12)$$

Solving for A and B we have

$$y_0(x) = y(x_0)\frac{u_1(x)}{u_1(x_0)}\left[1 + \left(\frac{y'(x_0)}{y(x_0)} - \frac{u_1'(x_0)}{u_1(x_0)}\right)\frac{1}{w_0(x_0)}\int_{x_0}^x w_0(x')dx'\right] \quad (4.13)$$

It is also useful to write this in terms of the Wronskian rather than $w_0(x)$, since the Wronskian can be expressed solely in terms of the coefficients $a_1(x)$ and $a_2(x)$ of the differential equation rather than the solution $u_1(x)$ (note (4.10)). We then have

$$y_0(x) = y(x_0)\frac{u_1(x)}{u_1(x_0)}\left[1 + \left(\frac{y'(x_0)}{y(x_0)} - \frac{u_1'(x_0)}{u_1(x_0)}\right)\frac{u_1^2(x_0)}{W(x_0)}\int_{x_0}^x \frac{W(x')}{u_1^2(x')}dx'\right]. \quad (4.14)$$

Next we consider the second order inhomogeneous equation

$$Ly(x) = a_2(x)y''(x) + a_1(x)y'(x) + a_0(x)y(x) = f(x) \quad (4.15)$$

and again assume that one solution, $u_1(x)$, to the homogeneous equation is known:

$$Lu_1(x) = 0. \quad (4.16)$$

Then, writing

$$y(x) = c(x)u_1(x), \quad (4.17)$$

the second degree equation $Ly(x) = f(x)$ is transformed into the first degree equation

$$\mathcal{L}(w) = w'(x)a_2(x)u_1(x) + w(x)\left(a_1(x)u_1(x) + 2a_2(x)u_1'(x)\right) = f(x) \quad (4.18)$$

where

$$w(x) = c'(x) \quad (4.19)$$

Writing (4.18) in the form

$$\left(\frac{w(x)}{w_0(x)}\right)' = \frac{f(x)}{a_2(x)u_1(x)w_0(x)} \quad (4.20)$$

where $w_0(x)$ is given by (4.8), a particular solution to the inhomogeneous equation is then, from (4.17), (4.19) and (4.20),

$$y_p(x) = u_1(x) \int^x w(x') \, dx' = u_1(x) \int^x dx' \, w_0(x') \int^{x'} \frac{f(x'')}{a_2(x'')u_1(x'')w_0(x'')} \, dx''$$
$$\text{(4.21)}$$

We note that if the lower limit of all the integrals in this expression for the particular solution are chosen to be x_0 then $y_p(x_0) = y_p'(x_0) = 0$. Writing y_p in terms of the Wronskian, we have

$$y_p(x) = u_1(x) \int_{x_0}^x dx' \, \frac{W(x')}{u_1^2(x')} \int_{x_0}^{x'} \frac{f(x'')u_1(x'')}{a_2(x'')W(x'')} \, dx'' \qquad \text{(4.22)}$$

A particular solution satisfying the initial conditions $y(x_0)$ and $y'(x_0)$ is obtained from (4.14) by adding $y_0(x)$ to $y_p(x)$, giving a solution to the inhomogeneous equation $Ly = f$ satisfying these initial conditions:

$$y(x) = u_1(x) \int_{x_0}^x dx' \, \frac{W(x')}{u_1^2(x')} \int_{x_0}^{x'} \frac{f(x'')u_1(x'')}{a_2(x'')W(x'')} \, dx''$$
$$+ y(x_0) \frac{u_1(x)}{u_1(x_0)} \left[1 + \left(\frac{y'(x_0)}{y(x_0)} - \frac{u_1'(x_0)}{u_1(x_0)} \right) \frac{u_1^2(x_0)}{W(x_0)} \int_{x_0}^x \frac{W(x')}{u_1^2(x')} \, dx' \right]. \qquad \text{(4.23)}$$

Turning now to the solution of the inhomogeneous equation given by the method of variation of constants, from (2.42) and (2.44), assuming that two linearly independent solutions of the homogeneous equation, $u_1(x)$ and $u_2(x)$, are known, we have

$$y(x) = u_2(x) \int_{x_0}^x \frac{f(x')u_1(x')}{a_2(x')W(x')} \, dx' - u_1(x) \int_{x_0}^x \frac{f(x')u_2(x')}{a_2(x')W(x')} \, dx' \qquad \text{(4.24)}$$

Here, for comparison with the solution given by reduction of order, we have chosen the lower limit of the integrals to be x_0. We note that the particular solutions given in both (4.22) and (4.24) satisfy $y(x_0) = y'(x_0) = 0$ and are therefore identical. This can be seen from (4.22) on integrating by parts. From (4.9) and (4.10), a second linearly independent solution to the homogeneous equation is

$$u_2(x) = u_1(x) \int_{x_0}^x w_0(x') \, dx' = u_1(x) \int_{x_0}^x \frac{W(x')}{u_1^2(x')} \, dx' \qquad \text{(4.25)}$$

Writing (4.22) in the form

$$y_p(x) = u_1(x) \int_{x_0}^x d\left(\int_{x_0}^{x'} \frac{W(x'')}{u_1^2(x'')} \, dx'' \right) \int_{x_0}^{x'} \frac{f(x'')u_1(x'')}{a_2(x'')W(x'')} \, dx'' \qquad \text{(4.26)}$$

and integrating by parts we have

$$
\begin{aligned}
y_p(x) &= \left(u_1(x) \int_{x_0}^x \frac{W(x')}{u_1^2(x')}\, dx' \right) \int_{x_0}^x \frac{f(x'')u_1(x'')}{a_2(x'')W(x'')}\, dx'' \\
&\quad - u_1(x) \int_{x_0}^x dx' \frac{f(x')u_1(x')}{a_2(x')W(x')} \int_{x_0}^{x'} \frac{W(x'')}{u_1^2(x'')}\, dx'' \\
&= u_2(x) \int_{x_0}^x \frac{f(x')u_1(x')}{a_2(x')W(x')}\, dx' - u_1(x) \int_{x_0}^x \frac{f(x')u_2(x')}{a_2(x')W(x')}\, dx'
\end{aligned}
\tag{4.27}
$$

from (4.25).

We note that, for a second order equation, although the methods of reduction of order and variation of constants give mathematically equivalent results, the resulting solutions are not in the same form: The solution using variation of constants is linear in the two functions $u_1(x)$ and $u_2(x)$, whereas the solution given by reduction of order is a non-linear function of $u_1(x)$. The analysis of inhomogeneous equations for which the solutions of the homogeneous equation are the special functions of mathematical physics may therefore be facilitated by choosing the method of variation of constants. This will be illustrated later with a specific example.

We next consider the method of reduction of order as given in (2.97)–(2.104) and apply it to the second order difference operator

$$
Ly(n) = p_2(n)y(n+2) + p_1(n)y(n+1) + p_0(n)y(n) \tag{4.28}
$$

We assume that one solution, $u_1(n)$, to the homogeneous equation is known:

$$
Lu_1(n) = 0 \tag{4.29}
$$

Then from (2.98), by writing

$$
y(n) = c(n)u_1(n) \tag{4.30}
$$

the second degree operator Ly is transformed into a first degree operator $\mathscr{L}w$ where

$$
w(n) = \Delta c(n) = c(n+1) - c(n) \tag{4.31}
$$

We denote by $w_0(n)$ the solution of $\mathscr{L}w = 0$, which then determines a second solution to the homogeneous equation $Ly = 0$:

$$
u_2(n) = c(n)u_1(n) = u_1(n) \sum_{k}^{n-1} w(k) \tag{4.32}
$$

From (2.104) with $N = 2$, the function $w_0(n)$ satisfies the first order homogeneous equation the first order equation for $w(n)$

$$p_2(n)u_1(n+2)\,\Delta w(n) + (p_1(n)u_1(n+1) + 2p_2(n)u_1(n+2))\,w(n) = 0 \tag{4.33}$$

which may be written in the form

$$p_2(n)u_1(n+2)w(n+1) + \big(p_1(n)u_1(n+1) + p_2(n)u_1(n+2)\big)w(n) = 0. \tag{4.34}$$

From (2.99) this equation may be written more simply as

$$p_2(n)u_1(n+2)w(n+1) = p_0(n)u_1(n)w(n). \tag{4.35}$$

Its solution follows by iteration, giving

$$w(n) = w(n_0)\frac{u_1(n_0)u_1(n_0+1)}{u_1(n)u_1(n+1)}\prod_{k=n_0}^{n-1}\frac{p_0(k)}{p_2(k)} \tag{4.36}$$

Here,

$$\begin{aligned} w(n_0)u_1(n_0)u_1(n_0+1) &= [c(n_0+1) - c(n_0)]u_1(n_0)u_1(n_0+1) \\ &= u_1(n_0)c(n_0+1)u_1(n_0+1) - u_1(n_0+1)c(n_0)u_1(n_0) \\ &= u_1(n_0)u_2(n_0+1) - u_1(n_0+1)u_2(n_0) \\ &= \mathcal{C}(n_0) \end{aligned} \tag{4.37}$$

where $\mathcal{C}(n)$ is the Casoratian determinant (see Appendix D). Thus,

$$w(n) = \frac{\mathcal{C}(n_0)}{u_1(n)u_1(n+1)}\prod_{k=n_0}^{n-1}\frac{p_0(k)}{p_2(k)} \tag{4.38}$$

from which, using (4.37),

$$\mathcal{C}(n) = \mathcal{C}(n_0)\prod_{k=n_0}^{n-1}\frac{p_0(k)}{p_2(k)}, \tag{4.39}$$

which is the equivalent of the equation for the Wronskian

$$W(x) = W(x_0)\exp\left(-\int_{x_0}^{x}\frac{a_1(x')}{a_2(x')}\,dx'\right) \tag{4.40}$$

for differential equations.

The function $c(n)$ then follows directly from (2.103) and (4.38) by summation, giving

$$c(n) = c(n_0) + \mathcal{C}(n_0) \sum_{j=n_0}^{n-1} \frac{1}{u_1(j)u_1(j+1)} \prod_{k=n_0}^{j-1} \frac{p_0(k)}{p_2(k)} \tag{4.41}$$

A second, linearly independent solution of the homogeneous equation is then

$$u_2(n) = c(n)u_1(n) = c(n_0)u_1(n) + u_1(n)\mathcal{C}(n_0) \sum_{j=n_0}^{n-1} \frac{1}{u_1(j)u_1(j+1)} \prod_{k=n_0}^{j-1} \frac{p_0(k)}{p_2(k)} \tag{4.42}$$

The first term on the right hand side here is simply a multiple of the first solution. For $N = 2$, the two linearly independent solutions of the second order homogeneous difference equation can thus be taken to be

$$u_1(n)$$

and

$$u_2(n) = u_1(n)\mathcal{C}(n_0) \sum_{j=n_0}^{n-1} \frac{1}{u_1(j)u_1(j+1)} \prod_{k=n_0}^{j-1} \frac{p_0(k)}{p_2(k)} \tag{4.43}$$

An alternate but completely equivalent approach to obtaining a second, linearly independent solution to a second order homogeneous equation when one solution is known is provided by consideration of the Wronskian (in the case of the differential equation) and the Casoratian (in the case of the difference equation).

For the second order homogeneous differential equation we have $n = 2$, so that from (C.1)

$$W(x) = u_1(x)u_2'(x) - u_1'(x)u_2(x) \tag{4.44}$$

which can be written as

$$\left(\frac{u_2(x)}{u_1(x)}\right)' = \frac{W(x)}{u_1^2(x)} \tag{4.45}$$

and integration gives

$$u_2(x) = u_1(x) \int^x \frac{W(x')}{u_1^2(x')} dx', \tag{4.46}$$

which is the solution given in (4.9).

For the second order homogeneous difference equation we have $N = 2$, so that from (D.1)

$$\mathcal{C}(n) = u_1(n)u_2(n+1) - u_1(n+1)u_2(n) \tag{4.47}$$

which can be written as

$$\frac{u_2(n+1)}{u_1(n+1)} - \frac{u_2(n)}{u_1(n)} = \frac{1}{u_1(n)u_1(n+1)}\mathcal{C}(n) \tag{4.48}$$

By summation we then have

$$\frac{u_2(n)}{u_1(n)} = \frac{u_2(0)}{u_1(0)} + \sum_{k=0}^{n-1} \frac{1}{u_1(k)u_1(k+1)}\mathcal{C}(k) \tag{4.49}$$

Substituting (D.3) in (4.49) we have

$$u_2(n) = \frac{u_2(0)}{u_1(0)}u_1(n) + u_1(n)\mathcal{C}(0) \sum_{k=0}^{n-1} \frac{1}{u_1(k)u_1(k+1)} \prod_{j=0}^{k-1} \frac{p_0(j)}{p_2(j)} \tag{4.50}$$

which is identical to the result obtained earlier in (4.42) noting that $u_2(0)/u_1(0) = c(0)$.

Chapter 5
Self-adjoint Linear Equations

As with most of the topics considered in this work, the concepts of adjoint and self-adjoint linear operators and equations, both differential and difference, apply to operators and equations of all orders. However, given the fundamental place of second order equations, both differential and difference, for problems in classical and quantum physics, we restrict ourselves here to equations of second order, noting that most of the classical functions of mathematical physics satisfy second order differential equations in the continuous variable and second order difference equations in the discrete variable. Self-adjoint operators, also called Hermitian operators, together with imposed boundary conditions, are of great importance in both classical and quantum physics within the framework of Sturm–Liouville theory, (note [3]), in that their eigenvalues are real, and their eigenfunctions are orthogonal and form a complete set. For analyses of higher order equations, see [38] and [20] for differential equations and [2] for difference equations.

The linear second order differential operator has the general form

$$Ly(x) \equiv a_2(x)y''(x) + a_1(x)y'(x) + a_0(x)y(x) \tag{5.1}$$

and the adjoint operator $\overline{L}$ is defined by

$$\overline{L}y(x) = (a_2(x)y(x))'' - (a_1(x)y(x))' + a_0(x)y(x) \tag{5.2}$$

Carrying out the differentiations in $\overline{L}$ and taking the adjoint again, we find the original operator L: The adjoint of the adjoint operator is the original operator.

If $a_2'(x) = a_1(x)$ then $\overline{L} = L$ and can be written in the form

$$\overline{L} = L = (a_2(x)y'(x))' + a_0(x)y(x) \tag{5.3}$$

The operator L is then said to be self-adjoint. We can, however, always transform a non-self-adjoint operator into the self-adjoint form: If $a_2'(x) \neq a_1(x)$, then

© Springer International Publishing Switzerland 2016
L.C. Maximon, *Differential and Difference Equations*,
DOI 10.1007/978-3-319-29736-1_5

multiplying $Ly(x)$ by a function $h(x)$, the resulting operator:

$$h(x)a_2(x)y''(x) + h(x)a_1(x)y'(x) + h(x)a_0(x)y(x) \qquad (5.4)$$

is then self-adjoint if we determine $h(x)$ by requiring that $(a_2(x)h(x))' = a_1(x)h(x)$. Solving for $h(x)$, we have

$$h(x) = \frac{1}{a_2(x)} \exp\left[\int_{x_0}^{x} \frac{a_1(x')}{a_2(x')} dx' \right] \qquad (5.5)$$

The linear second order difference operator can be written in the general form

$$Ly(n) = p_2(n)y(n+1) + p_1(n)y(n) + p_0(n)y(n-1) \qquad (5.6)$$

and the adjoint operator is defined by

$$\overline{L}y(n) = p_0(n+1)y(n+1) + p_1(n)y(n) + p_2(n-1)y(n-1) \qquad (5.7)$$

As with the differential operator, we note that the adjoint of the adjoint operator is the original operator.

If $p_2(n) = p_0(n+1)$ then $\overline{L} = L$ and can be written in the form

$$\overline{L} = L = \Delta\big(p_2(n-1)\Delta y(n-1)\big) + (p_0(n) + p_1(n) + p_2(n))y(n) \qquad (5.8)$$

The operator L is then said to be self-adjoint. We can, however, always transform a non-self-adjoint operator into the self-adjoint form: If $p_2(n) \neq p_0(n+1)$, then multiplying $Ly(n)$ by a function $h(n)$, the resulting operator:

$$h(n)p_2(n)y(n+1) + h(n)p_1(n)y(n) + h(n)p_0(n)y(n-1) \qquad (5.9)$$

is then self-adjoint if we determine $h(n)$ by requiring that $p_2(n)h(n) = p_0(n+1)h(n+1)$, that is,

$$h(n+1) = \frac{p_2(n)}{p_0(n+1)}h(n) \qquad (5.10)$$

Solving for $h(n)$, we have

$$h(n) = \prod_{j=n_0}^{n-1} \frac{p_2(j)}{p_0(j+1)}h(n_0+1) \qquad (5.11)$$

We assume here that $p_2(n) > 0$ and $p_0(n) > 0$ over the interval considered for n. The constant $h(n_0 + 1)$ is arbitrary and may be dropped.

The second order linear self-adjoint differential equation has the form

$$[P(x)z'(x)]' + Q(x)z(x) = 0 \tag{5.12}$$

The corresponding second order linear difference equation is

$$\Delta(p_{n-1}\Delta y_{n-1}) + q_n y_n = 0, \tag{5.13}$$

which may be derived from the self-adjoint differential equation as follows:

If the range of x is $c \le x \le d$ then we may picture this interval as a grid divided into N steps (N large), the individual steps being of length ϵ, where $\epsilon = (c - d)/N$. (If the range of x is infinite, for example $x \ge 0$, then we can define $\epsilon = 1/N$). A given point x on the grid is then associated with the integer n: $x = c + n\epsilon$. We can then write, for large N,

$$\begin{aligned}
\epsilon^2[P(x)z'(x)]' &= \epsilon^2 P'(x)z'(x) + \epsilon^2 P(x)z''(x) \\
&= [P(x) + \epsilon P(x)' + \tfrac{1}{2}\epsilon^2 P''(x)][(z(x) + \epsilon z(x)' + \tfrac{1}{2}\epsilon^2 z''(x)) - z(x)] \\
&\quad - P(x)[z(x) - (z(x) + \epsilon z(x)' + \tfrac{1}{2}\epsilon^2 z''(x))] + O(\epsilon^3) \\
&= P(x + \epsilon)[z(x + \epsilon) - z(x)] - P(x)[z(x) - z(x + \epsilon)] + O(\epsilon^3)
\end{aligned} \tag{5.14}$$

Writing

$$\begin{aligned}
x &= c + n\epsilon \\
y_n &\equiv z(c + n\epsilon) \\
p_{n-1} &\equiv P(c + n\epsilon) \\
q_n &\equiv \epsilon^2 Q(c + n\epsilon)
\end{aligned} \tag{5.15}$$

we then have

$$\begin{aligned}
\epsilon^2[P(x)z'(x)]' &= p_n(y_{n+1} - y_n) - p_{n-1}(y_n - y_{n-1}) + O(\epsilon^3) \\
&= p_n \Delta y_n - p_{n-1}\Delta y_{n-1} + O(\epsilon^3) \\
&= \Delta(p_{n-1}\Delta y_{n-1}) + O(\epsilon^3)
\end{aligned} \tag{5.16}$$

and

$$\epsilon^2\big[[P(x)z'(x)]' + Q(x)z(x)\big] = \Delta(p_{n-1}\Delta y_{n-1}) + q_n y_n + O(\epsilon^3). \tag{5.17}$$

Thus from (5.12),

$$\Delta(p_{n-1}\Delta y_{n-1}) + q_n y_n + O(\epsilon^3) = 0 \tag{5.18}$$

Here the first two terms are $O(\epsilon^2)$. Thus if we neglect terms of relative order ϵ, we obtain the second order linear difference equation

$$\Delta(p_{n-1}\Delta y_{n-1}) + q_n y_n = 0 \tag{5.19}$$

Chapter 6
Green's Function

6.1 Differential Equations

In Chap. 2, Sect. 2.1.1, we considered one method, variation of parameters (or variation of constants), for solving the linear inhomogeneous differential equation

$$Ly = f \tag{6.1}$$

where L is the differential operator

$$L \equiv a_n \frac{d^n}{dx^n} + a_{n-1} \frac{d^{n-1}}{dx^{n-1}} + \cdots + a_0 \tag{6.2}$$

and y and f are functions of x: $y = y(x)$, $f = f(x)$, the solution being considered in some interval, finite or infinite. In the method considered here, rather than determining the solution to (6.1) with the inhomogeneous term $f(x)$, which is defined at each point of the interval, we consider this equation when the inhomogeneous term is the Dirac delta function[1] $\delta(x - \xi)$, which is zero except at the point $x = \xi$ within the interval. Its solution is called the Green's function[2] $G(x, \xi)$:

$$LG(x, \xi) = \delta(x - \xi) \tag{6.3}$$

Further, since L is a linear operator (see Appendix F), we can weight the delta function with the factor $f(\xi)$ and integrate the contributions from all points of the interval to obtain the solution for the inhomogeneous term $f(x)$[3]:

---

[1] See [36, Sect. 1.17(i)].

[2] Named for the British mathematician George Green.

[3] Note that L is a function of x and hence can be taken outside the integral over ξ.

© Springer International Publishing Switzerland 2016
L.C. Maximon, *Differential and Difference Equations*,
DOI 10.1007/978-3-319-29736-1_6

$$L \int G(x, \xi) f(\xi) \, d\xi = \int \delta(x - \xi) f(\xi) \, d\xi = f(x) \qquad (6.4)$$

Then from (6.1) we have

$$Ly(x) - L \int G(x, \xi) f(\xi) \, d\xi \qquad (6.5)$$

from which

$$y(x) = \int G(x, \xi) f(\xi) \, d\xi \qquad (6.6)$$

We can also look at this result as defining the inverse of L, which expresses the solution to the differential equation (6.1) by writing

$$y = L^{-1} f \qquad (6.7)$$

where L^{-1} is the inverse of L, which is an integral operator in which the kernel is the Green's function, $G(x, \xi)$.

An important aspect of the Green's function is that it allows us to derive a solution $y(x)$ satisfying specific initial or boundary conditions by imposing those conditions on the Green's function $G(x, \xi)$; a brief discussion of these conditions follows.

The general solution to the nth order inhomogeneous equation consists of the sum of a particular solution, $y_p(x)$, and the n linearly independent solutions of the homogeneous equation $Ly(x) = 0$:

$$y(x) = y_p(x) + \sum_{k=1}^{n} c_k u_k(x) \qquad (6.8)$$

where

$$Lu_k(x) = 0 \qquad (6.9)$$

A solution satisfying specific initial or boundary conditions then involves the n constants $c_1, c_2 \ldots c_n$, which can be determined to satisfy the initial or boundary conditions. In an initial value problem, the unique solution of $Ly = f$ is determined by the value of the function $y(x)$ and its first $n - 1$ derivatives at the point $x = a$:

$$y(a) = \gamma_1, \ y^{(1)}(a) = \gamma_2, \ldots y^{(n-1)}(a) = \gamma_n \qquad (6.10)$$

These n initial conditions then determine the n constants $c_1, c_2 \ldots c_n$: From (6.8) we have n equations

$$y^{(j)}(a) = y_p^{(j)}(a) + \sum_{k=1}^{n} c_k u_k^{(j)}(a) = \gamma_{j+1}, \qquad j = 0, 1, \ldots n - 1 \qquad (6.11)$$

which determine the n constants c_k. By contrast, for a boundary value problem the n conditions are imposed at the two points a and b which define the boundary of the interval under consideration, $a \leq x \leq b$. Again, for the nth order equation there are n equations connecting the values of $y(x)$ and its first $n - 1$ derivatives at $x = a$ and $x = b$:

$$B_1 y = \alpha_{11} y(a) + \cdots + \alpha_{1n} y^{(n-1)}(a) + \beta_{11} y(b) + \cdots \beta_{1n} y^{(n-1)}(b) = \gamma_1$$

$$\vdots \tag{6.12}$$

$$B_n y = \alpha_{n1} y(a) + \cdots + \alpha_{nn} y^{(n-1)}(a) + \beta_{n1} y(b) + \cdots \beta_{nn} y^{(n-1)}(b) = \gamma_n$$

It should be noted that this set of boundary conditions includes the initial value conditions given above: If we let $\alpha_{11} = \alpha_{22} = \cdots = \alpha_{nn} = 1$, $\alpha_{jk} = 0$ if $j \neq k$, and all $\beta_{jk} = 0$, then we have the conditions given above for the initial value problem. There is, however, an important difference between the solution that is determined by the initial value conditions and the solution determined by the boundary value conditions. While the initial value conditions, imposed at one point, define a unique solution to the inhomogeneous equation, in the case of the boundary value conditions, imposed at two points, the inhomogeneous equation may have either a unique solution, many solutions, or no solution. An analysis of these possibilities with an example for a second order equation is presented in Appendix F.

The case of homogeneous boundary conditions: $B_k y = \gamma_k = 0$, $(k = 1, 2, \ldots n)$ is of particular importance in the analysis: From (6.6) we have

$$B_k y = \int_a^b B_k G(x, \xi) f(\xi) d\xi, \tag{6.13}$$

Thus if the Green's function is chosen to satisfy the homogeneous boundary conditions, then the solution $y(x)$ will also satisfy these homogeneous boundary conditions. The case of inhomogeneous boundary conditions ($\gamma_k \neq 0$) may then be treated using the superposition principle, discussed in Appendix F, by adding to the solution of the inhomogeneous equation ($f(x) \neq 0$) with homogeneous boundary conditions ($\gamma_k = 0$) a solution to the homogeneous equation ($f(x) = 0$) with inhomogeneous boundary conditions ($\gamma_k \neq 0$).

We turn now to the derivation of explicit expressions for the Green's function. We first present the details for the second order differential equation. This is followed by the analysis for the nth order equation, which is then fairly straightforward. As generally presented, the Green's function $G(x, \xi)$ for the second order equation is *constructed* to have the following properties:

(1)

$$G(x, s) = \begin{cases} G_1(x, \xi) & a \leq x < \xi \\ G_2(x, \xi) & \xi < x \leq b \end{cases} \tag{6.14}$$

where $LG_1(x, \xi) = 0$, $LG_2(x, \xi) = 0$.

(2) $G_1(x, \xi)$ and $G_2(x, \xi)$ satisfy the homogeneous boundary conditions $B_1 G_1 = B_2 G_2 = 0$.

(3) $G(x, \xi)$ is continuous at $x = \xi$; that is,

$$\lim_{\epsilon \to 0} [G_2(\xi + \epsilon, \xi) - G_1(\xi - \epsilon, \xi)] = 0 \qquad (6.15)$$

(4) The derivative $G'(x, \xi)$ has a discontinuity of magnitude $1/a_2(\xi)$ at $x = \xi$; that is,

$$\lim_{\epsilon \to 0} \left[G_2'(\xi + \epsilon, \xi) - G_1'(\xi - \epsilon, \xi) \right] = \frac{1}{a_2(\xi)} \qquad (6.16)$$

Here, however, we choose to view the Green's function in the light of our previous analysis of the method of variation of constants (see Chap. 2, Sect. 2.1.1). We therefore return to the inhomogeneous equation (6.3), $LG(x, \xi) = \delta(x - \xi)$, with $n = 2$. Assuming that the coefficients $a_0(x)$, $a_1(x)$ and $a_2(x)$ in the second order equation are continuous functions of x and that $a_2(x) \neq 0$ in $a \leq x \leq b$, we can write, from (6.6) and (6.3),

$$\int_{\xi - \epsilon}^{\xi + \epsilon} \left[\frac{d^2 G}{dx^2} + \frac{a_1(x)}{a_2(x)} \frac{dG}{dx} + \frac{a_0(x)}{a_2(x)} \right] dx = \int_{\xi - \epsilon}^{\xi + \epsilon} \frac{\delta(x - \xi)}{a_2(x)} dx \qquad (6.17)$$

Integrating both sides of this equation then gives

$$\frac{dG}{dx} \bigg|_{\xi - \epsilon}^{\xi + \epsilon} + \frac{a_1(\xi)}{a_2(\xi)} G \bigg|_{\xi - \epsilon}^{\xi + \epsilon} + 2\epsilon \frac{a_0(\xi)}{a_2(\xi)} = \frac{1}{a_2(\xi)} \qquad (6.18)$$

Assuming that $G(x, \xi)$ is continuous at $x = \xi$ and taking the limit $\epsilon \to 0$, we obtain the essential property of the Green's function, namely,

$$\lim_{\epsilon \to 0} \left[G'(\xi + \epsilon, \xi) - G'(\xi - \epsilon, \xi) \right] = \frac{1}{a_2(\xi)} \qquad (6.19)$$

Applying the method of variation of constants, we assume that we know two linearly independent solutions of the homogeneous equation, $u_1(x)$ and $u_2(x)$:

$$Lu_1(x) = 0 \qquad\qquad Lu_2(x) = 0 \qquad (6.20)$$

Following the analysis presented in Chap. 2, Sect. 2.1.1, noting (2.12), we write $G(x, \xi)$ and $G'(x, \xi)$ in terms of the two solutions of the homogeneous equation:

$$\begin{aligned} G(x, \xi) &= c_1(x, \xi) u_1(x) + c_2(x, \xi) u_2(x) \\ G'(x, \xi) &= c_1(x, \xi) u_1'(x) + c_2(x, \xi) u_2'(x) \end{aligned} \qquad (6.21)$$

where the parameters $c_1(x, \xi)$ and $c_2(x, \xi)$ now depend not only on x, as in Chap. 2, Sect. 2.1.1, but on the variable ξ as well. Differentiating the first equation in (6.21) we can also write

$$G'(x, \xi) = c_1(x, \xi)u_1'(x) + c_2(x, \xi)u_2'(x) + c_1'(x, \xi)u_1(x) + c_2'(x, \xi)u_2(x) \tag{6.22}$$

so that from (6.21) and (6.22)

$$c_1'(x, \xi)u_1(x) + c_2'(x, \xi)u_2(x) = 0 \tag{6.23}$$

Differentiation of the second equation in (6.21) gives

$$G'(x, \xi) = c_1(x, \xi)u_1''(x) + c_2(x, \xi)u_2''(x) + c_1'(x, \xi)u_1'(x) + c_2'(x, \xi)u_2'(x) \tag{6.24}$$

Finally, substituting (6.21) and (6.24) in (6.3), using (6.20), we have

$$c_1'(x, \xi)u_1'(x) + c_2'(x, \xi)u_2'(x) = \frac{\delta(x - \xi)}{a_2(x)} \tag{6.25}$$

From (6.23) and (6.25) we now have two equations for the first derivatives of the two parameters, $c_1(x, \xi)$ and $c_2(x, \xi)$, from which

$$\begin{aligned} c_1'(x, \xi) &= -\frac{u_2(x)\delta(x - \xi)}{a_2(x)W(x)} \\ c_2'(x, \xi) &= \frac{u_1(x)\delta(x - \xi)}{a_2(x)W(x)} \end{aligned} \tag{6.26}$$

where $W(x)$ is the Wronskian determinant:

$$W(x) = u_1(x)u_2'(x) - u_2(x)u_1'(x) \tag{6.27}$$

The integration of $c_1'(x, \xi)$ and $c_2'(x, \xi)$ introduces constants of integration which are constant with respect to x but may be functions of ξ:

$$c_k(x, \xi) = \int^x c_k'(x', \xi)dx' + \mu_k(\xi) \qquad k = 1, 2 \tag{6.28}$$

Further, we are free to choose the limit on these integrals; the particular choice will be one which simplifies satisfying the boundary conditions. To this end, reasonable choices are

$$\int_a^x c_k'(x', \xi)dx' \qquad \text{and} \qquad \int_x^b c_k'(x', \xi)dx' \tag{6.29}$$

For example, for the initial value problem with homogeneous initial value conditions $(y(a) = y'(a) = 0)$, we can set $\mu_1(\xi) = \mu_2(\xi) = 0$ and, from (6.26),

$$c_1(x, \xi) = -\int_a^x \frac{u_2(x')\delta(x'-\xi)}{a_2(x')W(x')} dx' = \begin{cases} -\dfrac{u_2(\xi)}{a_2(\xi)W(\xi)} & a < \xi < x \\ 0 & a \le x < \xi \end{cases} \tag{6.30}$$

$$c_2(x, \xi) = \int_a^x \frac{u_1(x')\delta(x'-\xi)}{a_2(x')W(x')} dx' = \begin{cases} \dfrac{u_1(\xi)}{a_2(\xi)W(\xi)} & a < \xi < x \\ 0 & a \le x < \xi \end{cases} \tag{6.31}$$

from which, from (6.21),

$$G(x, \xi) = \begin{cases} \dfrac{u_2(x)u_1(\xi) - u_1(x)u_2(\xi)}{a_2(\xi)W(\xi)} & a < \xi < x \\ 0 & a \le x < \xi \end{cases} \tag{6.32}$$

and, from (6.6), the solution to $Ly = f$ with homogeneous initial conditions imposed at $x = a$ is

$$y(x) = \int_a^x G(x, \xi) f(\xi) d\xi$$

$$= u_2(x) \int_a^x \frac{u_1(\xi)}{a_2(\xi)W(\xi)} f(\xi) d\xi - u_1(x) \int_a^x \frac{u_2(\xi)}{a_2(\xi)W(\xi)} f(\xi) d\xi \tag{6.33}$$

Note that it follows directly that $y(a) = 0$ and $y'(a) = 0$. If now we choose $u_1(x)$- and $u_2(x)$ to be linearly independent solutions of the homogeneous equation $Ly = 0$ satisfying the initial conditions $u_1(a) = 1$, $u_1'(a) = 0$ and $u_2(a) = 0$, $u_2'(a) = 1$, then $v(x) \equiv \gamma_1 u_1(x) + \gamma_2 u_2(x)$ is a solution of $Lv = 0$ with the initial conditions $v(a) = \gamma_1$, $v'(a) = \gamma_2$. The solution to $Ly = f$ with initial conditions $y(a) = \gamma_1$, $y'(a) = \gamma_2$ is therefore

$$y(x) = u_2(x) \int_a^x \frac{u_1(\xi)}{a_2(\xi)W(\xi)} f(\xi) d\xi - u_1(x) \int_a^x \frac{u_2(\xi)}{a_2(\xi)W(\xi)} f(\xi) d\xi$$
$$+ \gamma_1 u_1(x) + \gamma_2 u_2(x) \tag{6.34}$$

We next consider conditions imposed at the other end of the interval, at $x = b$, which we will call terminal conditions, namely, $y(b) = 0$, $y'(b) = 0$. We then choose, for the parameters $c_1(x, \xi)$ and $c_2(x, \xi)$,

$$c_1(x, \xi) = \int_x^b \frac{u_2(x')\delta(x'-\xi)}{a_2(x')W(x')} dx' = \begin{cases} 0 & \xi < x \le b \\ \dfrac{u_2(\xi)}{a_2(\xi)W(\xi)} & x < \xi < b \end{cases} \tag{6.35}$$

$$c_2(x, \xi) = -\int_x^b \frac{u_1(x')\delta(x' - \xi)}{a_2(x')W(x')} dx' = \begin{cases} 0 & \xi < x \le b \\ -\dfrac{u_1(\xi)}{a_2(\xi)W(\xi)} & x < \xi < b \end{cases} \quad (6.36)$$

from which

$$G(x, \xi) = \begin{cases} 0 & \xi < x \le b \\ \dfrac{u_1(x)u_2(\xi) - u_2(x)u_1(\xi)}{a_2(\xi)W(\xi)} & x < \xi < b \end{cases} \quad (6.37)$$

and, from (6.6), the solution to $Ly = f$ with homogeneous terminal conditions imposed at $x = b$ is

$$y(x) = \int_x^b G(x, \xi)f(\xi)d\xi$$

$$= u_1(x)\int_x^b \frac{u_2(\xi)}{a_2(\xi)W(\xi)} f(\xi)d\xi - u_2(x)\int_x^b \frac{u_1(\xi)}{a_2(\xi)W(\xi)} f(\xi)d\xi \quad (6.38)$$

If now we choose $u_1(x)$ and $u_2(x)$ to be linearly independent solutions of the homogeneous equation $Ly = 0$ satisfying the terminal conditions $u_1(b) = 1$, $u_1'(b) = 0$ and $u_2(b) = 0$, $u_2'(b) = 1$, then $v(x) \equiv \gamma_1 u_1(x) + \gamma_2 u_2(x)$ is a solution of $Lv = 0$ with the terminal conditions $v(b) = \gamma_1$, $v'(b) = \gamma_2$. The solution to $Ly = f$ with terminal conditions $y(b) = \gamma_1$, $y'(b) = \gamma_2$ is therefore

$$y(x) = u_1(x)\int_x^b \frac{u_2(\xi)}{a_2(\xi)W(\xi)} f(\xi)d\xi$$

$$- u_2(x)\int_x^b \frac{u_1(\xi)}{a_2(\xi)W(\xi)} f(\xi)d\xi + \gamma_1 u_1(x) + \gamma_2 u_2(x) \quad (6.39)$$

We next consider the boundary value problem in which each condition relates to only one end point:

$$\begin{aligned} B_1 y &= \alpha_{11} y(a) + \alpha_{12} y'(a) = \gamma_1 \\ B_2 y &= \beta_{21} y(b) + \beta_{22} y'(b) = \gamma_2 \end{aligned} \quad (6.40)$$

As before, $u_1(x)$ and $u_2(x)$ are linearly independent solutions of $Lu = 0$.

Again, we first consider the solution to the inhomogeneous equation $Ly = f$ with the homogeneous boundary conditions:

$$\begin{aligned} B_1 y &= \alpha_{11} y(a) + \alpha_{12} y'(a) = 0 \\ B_2 y &= \beta_{21} y(b) + \beta_{22} y'(b) = 0 \end{aligned} \quad (6.41)$$

In this case we choose

$$c_1(x, \xi) = \int_x^b \frac{u_2(x')\delta(x' - \xi)}{a_2(x')W(x')}dx' = \begin{cases} 0 & a < \xi < x \le b \\ \dfrac{u_2(\xi)}{a_2(\xi)W(\xi)} & a \le x < \xi < b \end{cases} \quad (6.42)$$

and

$$c_2(x, \xi) = \int_a^x \frac{u_1(x')\delta(x' - \xi)}{a_2(x')W(x')}dx' = \begin{cases} \dfrac{u_1(\xi)}{a_2(\xi)W(\xi)} & a < \xi < x \le b \\ 0 & a \le x < \xi < b \end{cases} \quad (6.43)$$

from which

$$G(x, \xi) = \begin{cases} \dfrac{u_2(x)u_1(\xi)}{a_2(\xi)W(\xi)} & a < \xi < x \le b \\ \dfrac{u_1(x)u_2(\xi)}{a_2(\xi)W(\xi)} & a \le x < \xi < b \end{cases} \quad (6.44)$$

and

$$y(x) = \int_a^b G(x, \xi)f(\xi)d\xi$$
$$= u_2(x)\int_a^x \frac{u_1(\xi)}{a_2(\xi)W(\xi)}f(\xi)d\xi + u_1(x)\int_x^b \frac{u_2(\xi)}{a_2(\xi)W(\xi)}f(\xi)d\xi \quad (6.45)$$

Therefore, if we choose $u_1(x)$ and $u_2(x)$ so that they satisfy the homogeneous boundary conditions

$$B_1u_1 = \alpha_{11}u_1(a) + \alpha_{12}u_1'(a) = 0$$
$$B_2u_2 = \beta_{21}u_2(b) + \beta_{22}u_2'(b) = 0 \quad (6.46)$$

(e.g., by setting $u_1(a) = \alpha_{12}, u'(a) = -\alpha_{11}, u_2(b) = \beta_{22}, u_2'(b) = -\beta_{21}$), then $G(x, \xi)$ as given in (6.44) will also satisfy the homogeneous boundary conditions (6.41) and hence $y(x)$ as given by (6.45) will also satisfy the homogeneous boundary conditions (6.41).

If now $y(x)$ is to satisfy the inhomogeneous boundary conditions (6.40), then we may add to $y(x)$ as given in (6.45) a solution to the homogeneous equation $Ly = 0$ satisfying the inhomogeneous boundary conditions (6.40), giving

$$y(x) = u_2(x)\int_a^x \frac{u_1(\xi)}{a_2(\xi)W(\xi)}f(\xi)d\xi + u_1(x)\int_x^b \frac{u_2(\xi)}{a_2(\xi)W(\xi)}f(\xi)d\xi$$
$$+ \mu_1 u_1(x) + \mu_2 u_2(x) \quad (6.47)$$

Imposing the inhomogeneous boundary conditions given in (6.40) and using (6.46), we have

$$B_1 y = \mu_2 B_1 u_2 = \gamma_1$$
$$B_2 y = \mu_1 B_2 u_1 = \gamma_2$$

(6.48)

from which

$$y(x) = u_2(x) \int_a^x \frac{u_1(\xi)}{a_2(\xi) W(\xi)} f(\xi) d\xi + u_1(x) \int_x^b \frac{u_2(\xi)}{a_2(\xi) W(\xi)} f(\xi) d\xi$$
$$+ \frac{\gamma_2}{B_2 u_1} u_1(x) + \frac{\gamma_1}{B_1 u_2} u_2(x)$$

(6.49)

We have derived three Green's functions, given in (6.32), (6.37) and (6.44), each satisfying a particular condition—initial, terminal, and boundary. It is important to note that each of these Green's functions is a solution of the inhomogeneous equation $LG(x, \xi) = \delta(x - \xi)$; they therefore differ only by a solution of the homogeneous equation $Lu = 0$. We can therefore use any one of these Green's functions in solving an initial, terminal or boundary value problem by adding an arbitrary solution to the homogeneous equation and then imposing the desired boundary conditions on the sum. For example, we could have dealt with the boundary value problem by adding to the Green's function for the initial value problem an arbitrary solution to the homogeneous equation and then imposing the boundary value conditions (6.41) on the sum:

$$G(x, \xi) = \mu_1(\xi) u_1(x) + \mu_2(\xi) u_2(x) + \begin{cases} \dfrac{u_2(x) u_1(\xi) - u_1(x) u_2(\xi)}{a_2(\xi) W(\xi)} & a < \xi < x \le b \\ 0 & a \le x < \xi < b \end{cases}$$

(6.50)

Imposing the boundary conditions (6.41) we have

$$B_1 G = \mu_1(\xi) B_1 u_1 + \mu_2(\xi) B_1 u_2 = 0$$
$$B_2 G = \mu_1(\xi) B_2 u_1 + \mu_2(\xi) B_2 u_2 + \frac{u_1(\xi) B_2 u_2 - u_2(\xi) B_2 u_1}{a_2(\xi) W(\xi)} = 0$$

(6.51)

Again, we assume that $u_1(x)$ and $u_2(x)$ satisfy the homogeneous boundary conditions (6.41)

$$B_1 u_1 = B_2 u_2 = 0$$

(6.52)

from which (see Appendix F) $B_1 u_2 \ne 0$, $B_2 u_1 \ne 0$ and hence

$$\mu_2(\xi) = 0$$
$$\mu_1(\xi) = \frac{u_2(\xi)}{a_2(\xi) W(\xi)}$$

(6.53)

from which we obtain the Green's function given in (6.44):

$$G(x, \xi) = \begin{cases} \dfrac{u_2(x)u_1(\xi)}{a_2(\xi)W(\xi)} & a < \xi < x \leq b \\[4mm] \dfrac{u_1(x)u_2(\xi)}{a_2(\xi)W(\xi)} & a \leq x < \xi < b \end{cases} \tag{6.54}$$

The observation that we can use any one of these Green's functions in solving an initial, terminal or boundary value problem by adding an arbitrary solution to the homogeneous equation and then imposing the desired boundary conditions on the sum is useful in considering the nth order equation, as will be seen. There we will choose the same limits on each of the integrals for the parameters $c_k(x, \xi)$.

Finally, we consider the solution of $Ly = f$ with the general boundary conditions involving the function and its first derivative at both end points of the interval, as given in (6.12). For the second order equation these boundary conditions are

$$\begin{aligned} B_1 y &= \alpha_{11} y(a) + \alpha_{12} y'(a) + \beta_{11} y(b) + \beta_{12} y'(b) = \gamma_1 \\ B_2 y &= \alpha_{21} y(a) + \alpha_{22} y'(a) + \beta_{21} y(b) + \beta_{22} y'(b) = \gamma_2 \end{aligned} \tag{6.55}$$

Again, we start by considering the solution to $Ly = f$ with homogeneous boundary conditions, (6.55) with $\gamma_1 = \gamma_2 = 0$). We note that in the initial, terminal and boundary value problems considered thus far, the respective Green's functions, (6.32), (6.37), and (6.44), each satisfied the homogeneous boundary conditions without the need to add the terms $\mu_1(\xi)u_1(x) + \mu_2(\xi)u_2(x)$ coming from the constants of integration. As a result, in the terms added to satisfy the inhomogeneous equation, the factors μ_1 and μ_2 were independent of ξ. Now, however, in the case of the general boundary conditions, in order to have the Green's function $G(x, \xi)$ satisfy the homogeneous boundary condition we have to write it in the form

$$G(x, \xi) = G_0(x, \xi) + \mu_1(\xi)u_1(x) + \mu_2(\xi)u_2(x) \tag{6.56}$$

where $G_0(x, \xi)$ can be chosen to be any one of the Green's functions given in (6.32), (6.37), and (6.44), all of which have the essential properties of being continuous at $x = \xi$ and having a discontinuous first derivative of magnitude $1/a_2(\xi)$ at $x = \xi$. Here, as before, $u_1(x)$ and $u_2(x)$ are any two linearly independent solutions of $Lu = 0$. We now impose the homogeneous boundary conditions on $G(x, \xi)$ and obtain the two equations which determine μ_1 and μ_2:

$$\begin{aligned} B_1 G &= B_1 G_0 + \mu_1 B_1 u_1 + \mu_2 B_1 u_2 = 0 \\ B_2 G &= B_2 G_0 + \mu_1 B_2 u_1 + \mu_2 B_2 u_2 = 0 \end{aligned} \tag{6.57}$$

from which

$$\mu_1 = \mu_1(\xi) = -\frac{1}{\Delta} \begin{vmatrix} B_1 G_0 & B_1 u_2 \\ B_2 G_0 & B_2 u_2 \end{vmatrix} \tag{6.58}$$

and

$$\mu_2 = \mu_2(\xi) = -\frac{1}{\Delta} \begin{vmatrix} B_1 u_1 & B_1 G_0 \\ B_2 u_1 & B_2 G_0 \end{vmatrix} \quad (6.59)$$

where

$$\Delta = \begin{vmatrix} B_1 u_1 & B_1 u_2 \\ B_2 u_1 & B_2 u_2 \end{vmatrix} \quad (6.60)$$

Recall from Appendix F that the condition $\Delta \neq 0$ is necessary and sufficient in order that $Ly = f$ with boundary conditions $B_1 y = B_2 y = 0$ have a unique solution. Substituting (6.58) and (6.59) in (6.56) then gives the Green's function which is a solution of $LG = \delta(x - \xi)$ and satisfies the homogeneous boundary conditions $B_1 G = B_2 G = 0$, in terms of which $y(x) = \int_a^b G(x, \xi) f(\xi) d\xi$ is a solution of $Ly = f$ and satisfies the homogeneous boundary conditions. The solution to $Ly = f$ with homogeneous boundary conditions is therefore

$$y(x) = \int_a^b G(x, \xi) f(\xi) d\xi$$
$$= \int_a^b \left[G_0(x, \xi) - \frac{u_1(x)}{\Delta} \begin{vmatrix} B_1 G_0 & B_1 u_2 \\ B_2 G_0 & B_2 u_2 \end{vmatrix} - \frac{u_2(x)}{\Delta} \begin{vmatrix} B_1 u_1 & B_1 G_0 \\ B_2 u_1 & B_2 G_0 \end{vmatrix} \right] f(\xi) d\xi$$
$$(6.61)$$

In order to have a solution to $Ly = f$ which satisfies the inhomogeneous boundary conditions (6.55), we may add a solution to the homogeneous equation $Lu = 0$ satisfying the inhomogeneous boundary conditions, viz., $v(x) = Au_1(x) + Bu_2(x)$ such that $B_1 v = \gamma_1$ and $B_2 v = \gamma_2$. These two conditions give two equations which determine the constants A and B, namely,

$$AB_1 u_1 + BB_1 u_2 = \gamma_1$$
$$AB_2 u_1 + BB_2 u_2 = \gamma_2 \quad (6.62)$$

from which

$$A = \frac{1}{\Delta} \begin{vmatrix} \gamma_1 & B_1 u_2 \\ \gamma_2 & B_2 u_2 \end{vmatrix}$$
$$B = \frac{1}{\Delta} \begin{vmatrix} B_1 u_1 & \gamma_1 \\ B_2 u_1 & \gamma_2 \end{vmatrix} \quad (6.63)$$

Adding $v(x) = Au_1(x) + Bu_2(x)$ to $y(x)$ as given in (6.61), the solution to $Ly = f$ with the inhomogeneous boundary conditions (6.55) is then

$$y(x) = \int_a^b \left[G_0(x, \xi) - \frac{u_1(x)}{\Delta} \begin{vmatrix} B_1 G_0 & B_1 u_2 \\ B_2 G_0 & B_2 u_2 \end{vmatrix} - \frac{u_2(x)}{\Delta} \begin{vmatrix} B_1 u_1 & B_1 G_0 \\ B_2 u_1 & B_2 G_0 \end{vmatrix} \right] f(\xi) d\xi$$

$$+ \frac{u_1(x)}{\Delta} \begin{vmatrix} \gamma_1 & B_1 u_2 \\ \gamma_2 & B_2 u_2 \end{vmatrix} + \frac{u_2(x)}{\Delta} \begin{vmatrix} B_1 u_1 & \gamma_1 \\ B_2 u_1 & \gamma_2 \end{vmatrix} \qquad (6.64)$$

It is worth summarizing with regard to this last expression: Here $y(x)$ satisfies the inhomogeneous differential equation $Ly = f$ and the inhomogeneous boundary conditions $B_1 y = \gamma_1$, $B_2 y = \gamma_2$. The only assumptions made with regard to the variables appearing in this expression are $LG_0 = \delta(x - \xi)$ and $Lu_1 = Lu_2 = 0$. The extension of this result to the nth order equation is straightforward: We now assume that $y(x)$ satisfies the nth order differential equation $Ly = f$ with L as given by (6.2), and the nth order general boundary conditions $B_k y = \gamma_k$, $(k = 1, 2, \ldots n)$, given in (6.12). Again, we first consider the solution of $Ly = f$ with the homogeneous boundary conditions $B_k y = 0$ and write the solution $y(x) = \int_a^b G(x, \xi) f(\xi) d\xi$ in terms of the Green's function

$$G(x, \xi) = G_0(x, \xi) + \sum_{k=1}^n \mu_k(\xi) u_k(x) \qquad (6.65)$$

where $LG_0(x, \xi) = \delta(x - \xi)$ and the n functions $u_k(x)$ are linearly independent solutions of $Lu = 0$. Solving $LG_0(x, \xi) = \delta(x - \xi)$ by the method of variation of constants as presented in Chap. 2, Sect. 2.1.1, we have, from (2.44) with $y(x)$ replaced by $G_0(x, \xi)$,

$$G_0(x, \xi) = \sum_{j=1}^n c_j(x, \xi) u_j(x), \qquad (6.66)$$

where $c_j(x, \xi) = \int^x c_j'(x', \xi) dx'$ in which $c_j'(x', \xi)$ is given by (2.42) with $g_n = \delta(x - \xi)/a_n(x)$. As noted before, the choice of the lower limit in this integral is arbitrary. For simplicity, we choose it to be $x = a$ for each of the n integrals, $j = 1, 2, \ldots n$. Integration then gives

$$c_j(x, \xi) = \begin{cases} \dfrac{(-1)^{n+j}}{a_n(\xi) W(\xi)} \begin{vmatrix} u_1(\xi) & \cdots & u_{j-1}(\xi) & u_{j+1}(\xi) & \cdots u_n(\xi) \\ u_1^{(1)}(\xi) & \cdots & u_{j-1}^{(1)}(\xi) & u_{j+1}^{(1)}(\xi) & \cdots u_n^{(1)}(\xi) \\ \vdots & \vdots & \vdots & \vdots & \vdots \\ u_1^{(n-2)}(\xi) & \cdots & u_{j-1}^{(n-2)}(\xi) & u_{j+1}^{(n-2)}(\xi) & \cdots u_n^{(n-2)}(\xi) \end{vmatrix} & a < \xi < x \le b \\[10pt] 0 & a \le x < \xi < b \end{cases}$$

$$(6.67)$$

(See Appendix B for $j = 1$ and $j = n$.) The Green's function $G_0(x, \xi)$ is then given by substituting $c_j(x, \xi)$ as given in (6.67) in (6.66). We note that for $n = 2$, (6.67) and (6.66) reduce directly to (6.30), (6.31) and (6.32). Imposing the homogeneous boundary conditions, (6.12) with $\gamma_k = 0$), on $G(x, \xi)$ as given in (6.65), we obtain n equations which determine the n coefficients $\mu_k(\xi)$, $(k = 1, 2, \ldots n)$:

$$B_k G = B_k G_0 + \sum_{j=1}^{n} \mu_j(\xi) B_k u_j = 0 \qquad (6.68)$$

from which, from Cramer's rule,

$$\mu_j(\xi) = -\frac{1}{\Delta} \begin{vmatrix} B_1 u_1 & \cdots & B_1 u_{j-1} & B_1 G_0 & B_1 u_{j+1} & \cdots & B_1 u_n \\ B_2 u_1 & \cdots & B_2 u_{j-1} & B_2 G_0 & B_2 u_{j+1} & \cdots & B_2 u_n \\ \vdots & & \vdots & \vdots & \vdots & & \vdots \\ B_n u_1 & \cdots & B_n u_{j-1} & B_n G_0 & B_n u_{j+1} & \cdots & B_n u_n \end{vmatrix} \qquad (6.69)$$

(See Appendix B for $j = 1$ and $j = n$.)
    Here

$$\Delta = \begin{vmatrix} B_1 u_1 & B_1 u_2 & \cdots & B_1 u_n \\ B_2 u_1 & B_2 u_2 & \cdots & B_2 u_n \\ \vdots & \vdots & & \vdots \\ B_n u_1 & B_n u_2 & \cdots & B_n u_n \end{vmatrix} \qquad (6.70)$$

(Note (6.60) for $n = 2$ and the discussion in Appendix F.) The solution to $Ly = f$ with homogeneous boundary conditions $B_k y = 0$ is then $y(x) = \int_a^b G(x, \xi) f(\xi) d\xi$ with $G(x, \xi)$ given by (6.65)–(6.67) and $\mu_k(\xi)$ given by (6.69) and (6.70).

    In order to have a solution to $Ly = f$ with the inhomogeneous boundary conditions $B_k y = \gamma_k$, we add a solution $u(x)$ which satisfies the homogeneous equation $Lu = 0$ and the inhomogeneous boundary conditions. This solution can be written in terms of the n linearly independent solutions of $Lu = 0$:

$$u(x) = \sum_{j=1}^{n} \eta_j u_j(x) \qquad (6.71)$$

Imposing the boundary conditions we have n equations which determine the n constants η_j:

$$\sum_{j=1}^{n} \eta_j B_k u_j = \gamma_k, \qquad k = 1, 2, \ldots, n \qquad (6.72)$$

from which

$$\eta_j = \frac{1}{\Delta} \begin{vmatrix} B_1 u_1 & \cdots & B_1 u_{j-1} & \gamma_1 & B_1 u_{j+1} & \cdots & B_1 u_n \\ B_2 u_1 & \cdots & B_2 u_{j-1} & \gamma_2 & B_2 u_{j+1} & \cdots & B_2 u_n \\ \vdots & & \vdots & \vdots & \vdots & & \vdots \\ B_n u_1 & \cdots & B_n u_{j-1} & \gamma_n & B_n u_{j+1} & \cdots & B_n u_n \end{vmatrix} \qquad (6.73)$$

(See Appendix B for $j = 1$ and $j = n$.)

The solution to the nth order equation $Ly = f$, (6.2), with general boundary conditions $B_k y = \gamma_k$, (6.12), is then

$$
\begin{aligned}
y(x) &= \int_a^b G(x, \xi) f(\xi) d\xi + \sum_{j=1}^n \eta_j u_j(x) \\
&= \int_a^b \left[G_0(x, \xi) + \sum_{k=1}^n \mu_k(\xi) u_k(x) \right] f(\xi) d\xi + \sum_{j=1}^n \eta_j u_j(x) \qquad (6.74)
\end{aligned}
$$

with $G_0(x, \xi)$ given in (6.66) and (6.67), $\mu_k(\xi)$ in (6.69), and η_j in (6.73).

Alternative analyses relevant to the nth order equation are given in [6, 7, 33, 38].

6.2 Difference Equations

In considering the Green's function for difference equations we follow closely the analysis just presented for differential equations. We start with the Nth order inhomogeneous difference equation, (2.45),

$$
\begin{aligned}
Ly(n) &= p_N(n)y(N + n) + p_{N-1}(n)y(N + n - 1) + \cdots + p_0(n)y(n) \\
&= \sum_{j=0}^N p_j(n)y(n + j) = q_N(n) \qquad (6.75)
\end{aligned}
$$

The Nth order homogeneous equation is

$$
Ly(n) = 0, \qquad (6.76)
$$

for which the N linearly independent solutions are denoted by $u_k(n)$, $(k = 1, 2, \ldots, N)$, that is,

$$
Lu_k(n) = \sum_{j=0}^N p_j(n)u_k(n + j) = 0, \qquad k = 1, 2, \ldots, N \qquad (6.77)
$$

Following the analysis presented for differential equations, we seek a solution to $Ly(n) = q(n)$ of the form

$$
y(n) = \sum_m G(n, m)q(m) \qquad (6.78)
$$

subject to certain initial, terminal, or boundary conditions.

As with the analysis of the differential equation, we first consider the second order difference equation in some detail and then generalize to the Nth order equation.

At this point a comment on the range of values for n and for m is in order. These ranges depend on whether the conditions are placed at the beginning of the interval, as in an initial value problem, at the end of the interval, or at both ends of the interval, as in a standard boundary value problem. We note that for the differential equation, both the function $y(x)$ *and* its derivative, $y'(x)$, were taken at the end points of the interval, at $x = a$ and $x = b$. For the difference equation the function is $y(n)$, but corresponding to the derivative we have $y(n + 1)$. Therefore, although we refer to $n = n_1$ and $n = n_2$ as the end points of the interval, $y(n)$, and hence also $G(n, m)$, are defined for $n_1 \leq n \leq n_2 + 1$.

In an initial value problem conditions are imposed on $y(n)$ at the beginning of the interval, on $y(n_1)$ and $y(n_1 + 1)$, the range of values of n is $n_1 \leq n < \infty$, and the range of values of m is $n_1 \leq m \leq n - 1$ (note (6.91) and (6.92)), the sum in (6.78) being zero for $n = n_1$. However, $y(n)$ and $G(n, m)$ are defined for all values of $n \geq n_1$.

In a terminal value problem conditions are imposed at the end of the interval, on $y(n_2)$ and $y(n_2 + 1)$ and n takes on the values $n_2 + 1, n_2, \ldots$ (i.e., $-\infty < n \leq n_2 + 1$) and the range of values of m is $n \leq m \leq n_2 - 1$ (note (6.102) and (6.103)), the sum in (6.78) being zero for $n = n_2$ and $n = n_2 + 1$. The functions $y(n)$ and $G(n, m)$ are defined for all values of n such that $-\infty < n \leq n_2 + 1$.

For a boundary value problem the conditions are imposed at both ends of the interval, on $y(n_1)$ and $y(n_1 + 1)$ and on $y(n_2)$ and $y(n_2 + 1)$ and the range of values of both n and m is $n_1 \leq n, m \leq n_2 - 1$ (note (6.108)). The functions $y(n)$ and $G(n, m)$ are defined for all values of n such that $n_1 \leq n \leq n_2 + 1$.

We therefore assume a solution to $Ly = q$ of the form (cf. (6.4))[4]

$$y(n) = \sum_{m=n_1}^{n_2-1} G(n, m)q(m) \tag{6.79}$$

We then have (cf. (6.5))

$$Ly = \sum_{m=n_1}^{n_2-1} LG(n, m)q(m) = q(n) \tag{6.80}$$

from which (noting that both n and m cover the range of values $n_1 \leq n, m \leq n_2 - 1$), (cf. (6.6))

$$LG(n, m) = \delta(n, m) \tag{6.81}$$

---

[4]To illustrate the similarity of the analyses, we give in italics the equation number in the corresponding derivation for differential equations.

where

$$\delta(n, m) = \begin{cases} 1 & n = m \\ 0 & n \neq m \end{cases} \tag{6.82}$$

As in the case of the differential equation, we solve the inhomogeneous difference equation (6.81) by the method of variation of constants, presented in Chap. 2,
Sect. 2.1.2, (cf. Eqs. (2.48) and (2.66) with $N = 2$). We then have (cf. (6.20))

$$G(n, m) = \sum_{k=1}^{2} c_k(n.m)u_k(n) \tag{6.83}$$

The coefficients $c_k(n, m)$ can be obtained by summing the differences $\Delta c_k(n' - 1, m)$
given in (2.82), in which $h_N(n - 1) = \delta(n - 1, m)/p_N(n - 1)$ (note Eq. (2.55)). To
this end, reasonable choices are (cf. (6.27) and (6.28))

$$c_k(n, m) = c_k(n_1, m) + \sum_{n'=n_1+1}^{n} \Delta c_k(n' - 1, m) \tag{6.84}$$

and

$$c_k(n, m) = c_k(n_2, m) - \sum_{n'=n+1}^{n_2} \Delta c_k(n' - 1, m) \tag{6.85}$$

Either of these expressions for $c_k(n, m)$ may be substituted in (6.83). The appropriate
choice permits one to satisfy the boundary conditions for $G(n, m)$ without adding
the solution to the homogeneous equation introduced in substituting the summation
constants $c_k(n_1, m)$ and $c_k(n_2, m)$ in (6.83). The coefficients $c_k(n, m)$ are then given
solely by the sums in (6.84) to (6.85). From (2.82) we then have (cf. (6.25))

$$\Delta c_1(n' - 1, m) = -\frac{\delta(n' - 1, m)}{p_2(n' - 1)K(n')}u_2(n')$$
$$\Delta c_2(n' - 1, m) = \frac{\delta(n' - 1, m)}{p_2(n' - 1)K(n')}u_1(n') \tag{6.86}$$

where from (2.56) (cf. (6.26))

$$K(n) = \begin{vmatrix} u_1(n) & u_2(n) \\ u_1(n + 1) & u_2(n + 1) \end{vmatrix} \tag{6.87}$$

Following the analysis presented in Eqs. (6.30)–(6.34) for differential equations,
we first consider the initial value problem with homogeneous initial value conditions $y(n_1) = y(n_1 + 1) = 0$. Then from (6.84) and (6.86) we choose (cf. (6.29) and
(6.30))

$$c_1(n, m) = - \sum_{n'=n_1+1}^{n} \frac{\delta(n'-1, m)}{p_2(n'-1)K(n')} u_2(n') = \begin{cases} -\dfrac{u_2(m+1)}{p_2(m)K(m+1)} & n_1 \le m < n \\ 0 & n_1 \le n \le m \end{cases}$$

(6.88)

$$c_2(n, m) = \sum_{n'=n_1+1}^{n} \frac{\delta(n'-1, m)}{p_2(n'-1)K(n')} u_1(n') = \begin{cases} \dfrac{u_1(m+1)}{p_2(m)K(m+1)} & n_1 \le m < n \\ 0 & n_1 \le n \le m \end{cases}$$

(6.89)

Note that we have chosen the interval of summation for both $c_1(n, m)$ and $c_2(n, m)$ to be $n_1 + 1 \le n' \le n$. This corresponds to (6.30) and (6.31), in which the interval of integration for both $c_1(x, \xi)$ and $c_2(x, \xi)$ was chosen to be $a < \xi < x$. This choice is the more reasonable one for an initial value problem.

It is worth noting that $c_1(n, m)$ and $c_2(n, m)$ appear to be functions only of m. Their n dependence specifies only the two regions of definition. This corresponds to the expressions for $c_1(x, \xi)$ and $c_2(x, \xi)$, which appear to be functions only of ξ.

Then from (6.83) we have (cf. (6.31))

$$G(n, m) = \begin{cases} \dfrac{u_2(n)u_1(m+1) - u_1(n)u_2(m+1)}{p_2(m)K(m+1)} & n_1 \le m < n \\ 0 & n_1 \le n \le m \end{cases}$$

(6.90)

It follows directly that $G(n_1, m) = G(n_1 + 1, m) = 0$ and hence from (6.79) the solution to $Ly = q(n)$ with the homogeneous initial conditions $y(n_1) = y(n_1 + 1) = 0$ is (cf. (6.32))

$$y(n) = \sum_{m=n_1}^{n-1} G(n, m)q(m)$$

$$= u_2(n) \sum_{m=n_1}^{n-1} \frac{u_1(m+1)}{p_2(m)K(m+1)} q(m) - u_1(n) \sum_{m=n_1}^{n-1} \frac{u_2(m+1)}{p_2(m)K(m+1)} q(m)$$

(6.91)

(Note (1.14) for $n = n_1$.)

If now we choose $u_1(n)$ and $u_2(n)$ to be linearly independent solutions of the homogeneous equation $Ly = 0$ satisfying the initial conditions $u_1(n_1) = 1$, $u_1(n_1 + 1) = 0$ and $u_2(n_1) = 0$, $u_2(n_1 + 1) = 1$, then $v(n) \equiv \gamma_1 u_1(n) + \gamma_2 u_2(n)$ is a solution of $Lv = 0$ with the initial conditions $v(n_1) = \gamma_1$, $v(n_1 + 1) = \gamma_2$. The solution to $Ly = q$ with initial conditions $y(n_1) = \gamma_1$, $y(n_1 + 1) = \gamma_2$ is therefore (cf. (6.33))

$$y(n) = u_2(n) \sum_{m=n_1}^{n-1} \frac{u_1(m+1)}{p_2(m)K(m+1)} q(m) - u_1(n) \sum_{m=n_1}^{n-1} \frac{u_2(m+1)}{p_2(m)K(m+1)} q(m)$$

$$+ \gamma_1 u_1(n) + \gamma_2 u_2(n) \tag{6.92}$$

(Note (6.34) for differential equations.)

It is interesting to compare the Green's function for differential equations given in (6.32) with that for difference equations given in (6.90). We note that from Eq. (6.90) (cf. (6.14))

$$G(m, m) = G(m+1, m) = 0, \tag{6.93}$$

which corresponds to the condition of continuity in (6.15), namely $\lim_{\epsilon \to 0} (G(\xi + \epsilon) - G(\xi - \epsilon)) = 0$. Further, from Eq. (6.90) (and (6.87)) we have (cf. (6.15))

$$G(m+2, m) = \frac{1}{p_2(m)}, \tag{6.94}$$

which corresponds to the discontinuity of $G'(x, \xi)$ at $x = \xi$, given in (6.16). From (6.90) and (2.47) it then follows that

$$LG(n, m) = p_2(n)G(n+2, m) + p_1(n)G(n+1, m) + p_0(n)G(n, m) = \delta(n, m) \tag{6.95}$$

which corresponds to $LG(x, \xi) = \delta(x - \xi)$, from which (cf. 6.5)

$$\sum_{n=m-1}^{n=m+1} LG(n, m) = 1 \tag{6.96}$$

which corresponds to

$$\int_{x=\xi-\epsilon}^{x=\xi+\epsilon} LG(x, \xi)\, dx = 1 \tag{6.97}$$

We next consider the case in which the boundary values are placed at the end of the interval, with homogeneous boundary conditions $y(n_2) = y(n_2 + 1) = 0$. Again, we choose from (6.84) and (6.85) the expressions for $c_k(n, m)$ that permit the boundary conditions for $G(n, m)$ to be satisfied without adding the solution to the homogeneous equation introduced in substituting the summation constants $c_k(n_1, m)$ and $c_k(n_2, m)$ in (6.83). The coefficients $c_k(n, m)$ are then given solely by the sums in (6.84) and (6.85). Then from (6.85) and (6.86) (corresponding to (6.35) and (6.36) for the differential equation),

$$c_1(n, m) = \sum_{n'=n+1}^{n_2} \frac{\delta(n'-1, m)}{p_2(n'-1)K(n')} u_2(n') = \begin{cases} \frac{u_2(m+1)}{p_2(m)K(m+1)} & n \le m \le n_2 - 1 \\ 0 & m < n \le n_2 - 1 \end{cases}$$

$$(6.98)$$

$$c_2(n, m) = -\sum_{n'=n+1}^{n_2} \frac{\delta(n'-1, m)}{p_2(n'-1)K(n')} u_1(n') = \begin{cases} -\frac{u_1(m+1)}{p_2(m)K(m+1)} & n \le m \le n_2 - 1 \\ 0 & m < n \le n_2 - 1 \end{cases}$$

$$(6.99)$$

and from (6.83) (cf. (6.36))

$$G(n, m) = \begin{cases} \dfrac{u_1(n)u_2(m+1) - u_2(n)u_1(m+1)}{p_2(m)K(m+1)} & n \le m \le n_2 - 1 \\ 0 & m < n \le n_2 - 1 \end{cases} \quad (6.100)$$

Note that this equation only defines $G(n, m)$ for $n \le n_2 - 1$. From (6.81) and (6.100) it follows, with a bit of algebra, that

$$G(n_2, m) = G(n_2 + 1, m) = 0 \quad \text{for all} \quad m \le n_2 - 1 \quad (6.101)$$

The solution to $Ly = q(n)$ with the homogeneous terminal conditions $y(n_2) = y(n_2 + 1) = 0$ is then (cf. (6.37))

$$y(n) = \sum_{m=n}^{n_2-1} G(n, m)q(m)$$

$$= u_1(n) \sum_{m=n}^{n_2-1} \frac{u_2(m+1)}{p_2(m)K(m+1)} q(m) - u_2(n) \sum_{m=n}^{n_2-1} \frac{u_1(m+1)}{p_2(m)K(m+1)} q(m)$$

$$(6.102)$$

(Note (1.14) for $n = n_2$ and $n = n_2 + 1$.)

If now we choose $u_1(n)$ and $u_2(n)$ to be linearly independent solutions of the homogeneous equation $Ly = 0$ satisfying the terminal conditions $u_1(n_2) = 1$, $u_1(n_2 + 1) = 0$ and $u_2(n_2) = 0$, $u_2(n_2 + 1) = 1$, then $v(n) \equiv \gamma_1 u_1(n) + \gamma_2 u_2(n)$ is a solution of $Lv = 0$ with the boundary conditions $v(n_2) = \gamma_1$, $v(n_2 + 1) = \gamma_2$. The solution to $Ly = q$ with boundary conditions $y(n_2) = \gamma_1$, $y(n_2 + 1) = \gamma_2$ is therefore (cf. (6.38))

$$y(n) = u_1(n) \sum_{m=n}^{n_2-1} \frac{u_2(m+1)}{p_2(m)K(m+1)} q(m) - u_2(n) \sum_{m=n}^{n_2-1} \frac{u_1(m+1)}{p_2(m)K(m+1)} q(m)$$

$$+ \gamma_1 u_1(n) + \gamma_2 u_2(n) \quad (6.103)$$

We next consider the boundary value problem with conditions imposed at both ends of the interval, but in which each condition relates to only one end point: (cf. (6.39))

$$B_1 y = \alpha_{11} y(n_1) + \alpha_{12} y(n_1 + 1) = \gamma_1$$
$$B_2 y = \beta_{21} y(n_2) + \beta_{22} y(n_2 + 1) = \gamma_2 \qquad (6.104)$$

Again, we first consider the solution to the inhomogeneous equation $Ly = q$ with the homogeneous boundary conditions: (cf. (6.40))

$$B_1 y = \alpha_{11} y(n_1) + \alpha_{12} y(n_1 + 1) = 0$$
$$B_2 y = \beta_{21} y(n_2) + \beta_{22} y(n_2 + 1) = 0 \qquad (6.105)$$

Then, corresponding to (6.42) and (6.43) for differential equations, we choose the summation interval $n + 1 \leq n' \leq n_2$ for $c_1(n, m)$ and $n_1 + 1 \leq n' \leq n$ for $c_2(n, m)$. Again, we choose from (6.84) and (6.85) the expressions for $c_k(n, m)$ that permit the boundary conditions for $G(n, m)$ to be satisfied without adding the solution to the homogeneous equation introduced in substituting the summation constants $c_k(n_1, m)$ and $c_k(n_2, m)$ in (6.83). The coefficients $c_k(n, m)$ are then given solely by the sums in (6.84) and (6.85). From (6.85) and (6.86) (corresponding to (6.35) for the differential equation), we then have (cf. (6.41))

$$c_1(n, m) = \sum_{n'=n+1}^{n_2} \frac{\delta(n' - 1, m)}{p_2(n' - 1) K(n')} u_2(n') = \begin{cases} \frac{u_2(m+1)}{p_2(m) K(m+1)} & n_1 \leq n \leq m \\ 0 & m < n \leq n_2 - 1 \end{cases}$$
$$(6.106)$$

and from (6.84) and (6.86) (corresponding to (6.31) for the differential equation), we have (cf. (6.42))

$$c_2(n, m) = \sum_{n'=n_1+1}^{n} \frac{\delta(n' - 1, m)}{p_2(n' - 1) K(n')} u_1(n') = \begin{cases} 0 & n_1 \leq n \leq m \\ \frac{u_1(m+1)}{p_2(m) K(m+1)} & m < n \leq n_2 - 1 \end{cases}$$
$$(6.107)$$

We then have (cf. (6.43))

$$G(n, m) = \begin{cases} \dfrac{u_1(n) u_2(m + 1)}{p_2(m) K(m + 1)} & n_1 \leq n \leq m \leq n_2 - 1 \\[4mm] \dfrac{u_2(n) u_1(m + 1)}{p_2(m) K(m + 1)} & n_1 \leq m < n \leq n_2 - 1 \end{cases} \qquad (6.108)$$

and (cf. (6.44))

$$y(n) = \sum_{m=n_1}^{n_2-1} G(n, m) q(m)$$

$$= u_2(n) \sum_{m=n_1}^{n-1} \frac{u_1(m + 1)}{p_2(m) K(m + 1)} q(m) + u_1(n) \sum_{m=n}^{n_2-1} \frac{u_2(m + 1)}{p_2(m) K(m + 1)} q(m)$$
$$(6.109)$$

If we choose $u_1(n)$ and $u_2(n)$ so that they satisfy the homogeneous boundary conditions (cf. (6.45))

$$B_1 u_1 = \alpha_{11} u_1(n_1) + \alpha_{12} u_1(n_1 + 1) = 0$$
$$B_2 u_2 = \beta_{21} u_2(n_2) + \beta_{22} u_2(n_2 + 1) = 0 \tag{6.110}$$

(e.g., by setting $u_1(n_1) = \alpha_{12}$, $u(n_1 + 1) = -\alpha_{11}$, $u_2(n_2) = \beta_{22}$, $u_2(n_2 + 1) = -\beta_{21}$) then $G(n, m)$ as given in (6.108) and $y(n)$ as given by (6.109) will also satisfy the homogeneous boundary conditions (6.105).

If now $y(n)$ is to satisfy the inhomogeneous boundary conditions (6.104), then we may add to $y(n)$ as given in (6.109) a solution to the homogeneous equation $Ly = 0$ satisfying the inhomogeneous boundary conditions (6.104), giving (cf. (6.46))

$$y(n) = u_2(n) \sum_{m=n_1}^{n-1} \frac{u_1(m+1)}{p_2(m) K(m+1)} q(m) + u_1(n) \sum_{m=n}^{n_2-1} \frac{u_2(m+1)}{p_2(m) K(m+1)} q(m)$$
$$+ \mu_1 u_1(n) + \mu_2 u_2(n) \tag{6.111}$$

Imposing the inhomogeneous boundary conditions given in (6.104) and using (6.110), we have (cf. (6.47))

$$B_1 y = \mu_2 B_1 u_2 = \gamma_1$$
$$B_2 y = \mu_1 B_2 u_1 = \gamma_2 \tag{6.112}$$

from which (cf. (6.48))

$$y(n) = u_2(n) \sum_{m=n_1}^{n-1} \frac{u_1(m+1)}{p_2(m) K(m+1)} q(m) + u_1(n) \sum_{m=n}^{n_2-1} \frac{u_2(m+1)}{p_2(m) K(m+1)} q(m)$$
$$+ \frac{\gamma_2}{B_2 u_1} u_1(x) + \frac{\gamma_1}{B_1 u_2} u_2(x) \tag{6.113}$$

We have derived three Green's functions, given in (6.90), (6.100) and (6.108), each satisfying a particular condition—initial, terminal, and boundary. It is important to note that each of these Green's functions is a solution of the inhomogeneous equation $LG(n, m) = \delta(n, m)$; they therefore only differ by a solution of the homogeneous equation $Lu = 0$. We can therefore use any one of these Green's functions in solving an initial, terminal or boundary value problem by adding an arbitrary solution to the homogeneous equation and then imposing the desired boundary conditions on the sum. For example, we could have dealt with the boundary value problem by adding to the Green's function for the initial value problem an arbitrary solution to the homogeneous equation and then imposing the boundary value conditions (6.41) on the sum: (cf. (6.49))

$$G(n, m) = \mu_1(m)u_1(n) + \mu_2(m)u_2(n) + \begin{cases} \frac{u_2(n)u_1(m+1)-u_1(n)u_2(m+1)}{p_2(m)K(m+1)} & n_1 \le m < n \\ 0 & n_1 \le n \le m \end{cases}$$
$$\text{(6.114)}$$

Imposing the boundary conditions (6.105) we have (cf. (6.50))

$$B_1 G = \mu_1(m)B_1 u_1 + \mu_2(m)B_1 u_2 = 0$$

$$B_2 G = \mu_1(m)B_2 u_1 + \mu_2(m)B_2 u_2 + \frac{u_1(m+1)B_2 u_2 - u_2(m+1)B_2 u_1}{p_2(m)K(m+1)} = 0$$
$$\text{(6.115)}$$

Again, we assume that $u_1(n)$ and $u_2(n)$ satisfy the homogeneous boundary conditions (6.110) (cf. (6.51))

$$B_1 u_1 = B_2 u_2 = 0 \tag{6.116}$$

from which (see Appendix F) $B_1 u_2 \ne 0$, $B_2 u_1 \ne 0$ and hence (cf. (6.52))

$$\mu_2(m) = 0$$
$$\mu_1(m) = \frac{u_2(m+1)}{p_2(m)K(m+1)} \tag{6.117}$$

from which we obtain the Green's function given in (6.108): (cf. (6.53))

$$G(n, m) = \begin{cases} \dfrac{u_1(n)u_2(m+1)}{p_2(m)K(m+1)} & n_1 \le n \le m \\ \dfrac{u_2(n)u_1(m+1)}{p_2(m)K(m+1)} & n_1 \le m < n \end{cases} \tag{6.118}$$

The observation that we can use any one of these Green's functions in solving an initial, terminal or boundary value problem by adding an arbitrary solution to the homogeneous equation and then imposing the desired boundary conditions on the sum is useful in considering the Nth order equation in that we can then choose the same limits on each of the sums for the parameters $c_k(n, m)$, just as, when considering the nth order differential equation, the limits on each of the integrals of $c'_j(x', \xi)$ were taken to be the same.

Finally, we consider the solution of $Ly = q$ with general boundary conditions involving the function at both ends of the interval, corresponding to (6.55) for differential equations. For the second order equation, these boundary conditions are (cf. (6.54))

$$B_1 y = \alpha_{11} y(n_1) + \alpha_{12} y(n_1 + 1) + \beta_{11} y(n_2) + \beta_{12} y(n_2 + 1) = \gamma_1$$
$$B_2 y = \alpha_{21} y(n_1) + \alpha_{22} y(n_1 + 1) + \beta_{21} y(n_2) + \beta_{22} y(n_2 + 1) = \gamma_2$$
$$\text{(6.119)}$$

Again, we start by considering the solution to $Ly = q$ with homogeneous boundary conditions, ((6.119) with $\gamma_1 = \gamma_2 = 0$). Following the analysis given in Eqs. (6.55)–(6.64) for differential equations, we obtain the Green's function $G(n, m)$ which satisfies the homogeneous boundary conditions by writing it in the form (cf. (6.55))

$$G(n, m) = G_0(n, m) + \mu_1(m)u_1(n) + \mu_2(m)u_2(n) \tag{6.120}$$

where, as we have discussed, $G_0(n, m)$ can be chosen to be any one of the Green's functions given in (6.90), (6.100) and (6.108). Imposing the general boundary conditions given in (6.119) (with $\gamma_1 = \gamma_2 = 0$) again leads to the two equations given in (6.57) for differential equations, from which the expressions for μ_1 and μ_2 given in (6.58)–(6.60) follow (with ξ replaced by m). The solution to $Ly = q$ with the general homogeneous boundary conditions is therefore (cf. (6.60))

$$y(n) = \sum_{m=n_1}^{n_2-1} G(n, m)q(m)$$

$$= \sum_{m=n_1}^{n_2-1} \left[G_0(n, m) - \frac{u_1(n)}{\Delta} \begin{vmatrix} B_1 G_0 & B_1 u_2 \\ B_2 G_0 & B_2 u_2 \end{vmatrix} - \frac{u_2(n)}{\Delta} \begin{vmatrix} B_1 u_1 & B_1 G_0 \\ B_2 u_1 & B_2 G_0 \end{vmatrix} \right] q(m) \tag{6.121}$$

The extension of these results to the Nth order difference equation follows in analogy with those for the nth order differential equation given in Eqs. (6.65)–(6.74). Now, in analogy with the analysis on p. 50, we consider the solution of $Ly = q$ with L as given in (6.75) and the Nth order general boundary conditions $B_k y = \gamma_k$, where

$$B_k y = \sum_{j=1}^{N} \left[\alpha_{kj} y(n_1 + j - 1) + \beta_{kj} y(n_2 + j - 1) \right], \quad k = 1, 2, \ldots, N \tag{6.122}$$

Again, we first consider the solution of $Ly = q$ with homogeneous boundary conditions, ($B_k y = 0$), and write the solution

$$y(n) = \sum_{m=n_1}^{n_2-1} G(n, m)q(m) \tag{6.123}$$

in terms of the Green's function

$$G(n, m) = G_0(n, m) + \sum_{k=1}^{N} \mu_k(m)u_k(n) \tag{6.124}$$

where

$$LG_0(n, m) = \delta(n - 1, m) \tag{6.125}$$

and the N functions $u_k(n)$ are linearly independent solutions of $Lu = 0$. Solving $LG_0(n, m) = \delta(n - 1, m)$ by the method of variation of constants as presented in Chap. 2, Sect. 2.1.1, we have, from (2.85) with $y(n)$ replaced by $G_0(n, m)$,

$$G_0(n, m) = \sum_{j=1}^{N} c_j(n, m) u_j(n), \tag{6.126}$$

where

$$c_j(n, m) = \sum_{n'=n_1+1}^{n} \Delta c_j(n' - 1, m) \tag{6.127}$$

in which $\Delta c_j(n' - 1, m)$ is given by (2.82) with $h_N(n - 1) = \delta(n - 1, m)/p_N(n - 1)$. Summation then gives

$$
c_j(n, m) =
\begin{cases}
\dfrac{(-1)^{N+j}}{p_N(m) K(m + 1)}
\begin{vmatrix}
u_1(m + 1) & \cdots & u_{j-1}(m + 1) & u_{j+1}(m + 1) & \cdots u_N(m + 1) \\
u_1(m + 2) & \cdots & u_{j-1}(m + 2) & u_{j+1}(m + 2) & \cdots u_N(m + 2) \\
\vdots & & \vdots & \vdots & \vdots \\
u_1(m + N - 1) & \cdots & u_{j-1}(m + N - 1) & u_{j+1}(m + N - 1) & \cdots u_N(m + N - 1)
\end{vmatrix} & n_1 \leq m < n \\[2em]
0 & n_1 \leq n \leq m
\end{cases}
\tag{6.128}
$$

(See Appendix B for $j = 1$ and $j = N$.) The Green's function $G_0(n, m)$ is then given by substituting $c_j(n, m)$ as given in (6.128) in (6.126). We note that for $N = 2$, (6.128) and (6.126) reduce directly to (6.88), (6.89) and (6.90).

The remaining analysis needed to express the solution to the difference equation $Ly = q$ in terms of the Green's function with general inhomogeneous boundary conditions is that given for differential equations in Eqs. (6.68)–(6.74); one has only to replace x and ξ by n and m, respectively, and apply the general boundary conditions $B_k y = \gamma_k$ given in (6.122). The solution to the Nth order difference equation $Ly = q$, (6.75), is then

$$
\begin{aligned}
y(n) &= \sum_{m=n_1}^{n_2-1} G(n, m) q(m) + \sum_{j=1}^{N} \eta_j u_j(n) \\
&= \sum_{m=n_1}^{n_2-1} \left[G_0(n, m) + \sum_{j=1}^{N} \mu_j(m) u_j(n) \right] q(m) + \sum_{j=1}^{N} \eta_j u_j(n)
\end{aligned}
\tag{6.129}
$$

Chapter 7
Generating Functions, Z-Transforms, Laplace Transforms and the Solution of Linear Differential and Difference Equations

Laplace transforms provide one of the means for solving homogeneous and inhomogeneous differential equations. Generating functions provide the corresponding transform for difference equations. The Laplace transform, $\mathcal{L}$, of a function $y(x)$ is defined by

$$\mathcal{L}y = \int_0^\infty e^{-sx} y(x)\, dx = F(s) \tag{7.1}$$

The generating function, $G(\omega)$, of a function $y(n)$ is defined by

$$\mathcal{G}y = \sum_{n=0}^\infty y(n)\omega^n = G(\omega) \tag{7.2}$$

A z-transform, also called a Laurent transform, is a generating function in which the variable ω is replaced by $z = 1/\omega$:

$$\mathcal{Z}y = \sum_{n=0}^\infty \frac{y(n)}{z^n} = Z(z) \tag{7.3}$$

The underlying similarity of these transforms may be seen from their definition and is made clear if, in the integrand of the Laplace transform, we consider a function $Y(x)$ defined as a sum of delta functions:

$$Y(x) = \sum_{n=0}^\infty \delta(x - n)y(x) \tag{7.4}$$

and let $e^{-s} = \omega$. Then

$$\mathcal{L}Y = \int_0^\infty e^{-sx} Y(x)\, dx = \sum_{n=0}^\infty \int_0^\infty \omega^x \delta(x - n)y(x)\, dx = \sum_{n=0}^\infty \omega^n y(n) \tag{7.5}$$

which is the generating function for the function $y(n)$.

© Springer International Publishing Switzerland 2016
L.C. Maximon, *Differential and Difference Equations*,
DOI 10.1007/978-3-319-29736-1_7

Both Laplace transforms and generating functions are generally applied to linear equations with constant coefficients since the resulting transform, $F(s)$ or $G(\omega)$, is then an algebraic function of the transform variable, s or ω, and the transform can therefore be inverted easily to give the solution to the differential or difference equation, $Ly = f(x)$ or $Ly = q(n)$. This case is presented in the next two sections. In the sections that follow we consider the case in which the coefficients in the differential or difference equation are polynomials (in x for differential equations, in n for difference equations). As we shall see, the transform, $F(s)$ or $G(\omega)$, is then no longer a simple algebraic expression, but now satisfies a differential equation whose order is equal to the highest degree of any of the polynomial coefficients. It is clear that this is not a reasonable method for the solution of the differential or difference equation unless the coefficients are either linear or quadratic functions of the independent variable, x or n. We therefore examine in some detail the case in which the coefficients are linear functions (of x for differential equations, of n for difference equations), with particular reference to some of the classical orthogonal polynomials. Nonetheless, obtaining the solution, $y(x)$ or $y(n)$, by inversion of the transform: $y(x) = \mathcal{L}^{-1} F(s)$ or $y(n) = \mathcal{G}^{-1} G(w)$ is, in general, somewhat difficult. We therefore present the approach generally taken, which consists in assuming that the solution to the differential or difference equation is given by an integral over the Laplace transform or generating function (the inverse Laplace or Mellin transform) and then using the equation, $Ly = f(x)$ or $Ly = q(n)$, to derive the necessary conditions that the Laplace transform or generating function must satisfy. (See, e.g., [25, Chap. 5, Sect. 47] and [27, A, Sect. 5, 19] in the case of differential equations and [26, Chap. XI, Sect. 174], or [35, Chap. 11] in the case of difference equations). Finally, for the case of a second order homogeneous equation we present an alternate approach in which the dependent and independent variables of the differential or difference equation are transformed so that the resulting equation is recognized to have a form whose solution is well-known. We apply this approach to the second order homogeneous differential equation with coefficients linear in the independent variable (for which details are given in [13, Sect. 6.2, pp. 249–252]) as well as to the corresponding difference equation.

7.1 Laplace Transforms and the Solution of Linear Differential Equations with Constant Coefficients

We first consider the case of a linear inhomogeneous differential equation with constant coefficients:

$$Ly = a_N y^{(N)}(x) + a_{N-1} y^{(N-1)}(x) + \cdots + a_0 y(x) = f(x) \qquad (7.6)$$

The solution to this equation will be obtained from its Laplace transform, defined by

$$F(s) = \mathcal{L}y(x) = \int_0^\infty e^{-sx} y(x) \, dx, \qquad (7.7)$$

from which $y(x)$ is given by the inverse Laplace transform:

$$y(x) = \mathcal{L}^{-1} F(s) \qquad (7.8)$$

Using the differential equation $Ly = f(x)$, we first obtain the Laplace transform of Ly and then express the Laplace transform of $y(x)$ in terms of Laplace transform of Ly.

To obtain the Laplace transform of Ly we define

$$F_k(s) = \int_0^\infty e^{-sx} y^{(k)}(x) \, dx \qquad k = 0, 1, \ldots N \qquad (F_0(s) = F(s)) \qquad (7.9)$$

from which, by integration by parts,

$$F_k(s) = s F_{k-1}(s) - y^{(k-1)}(0) \qquad k = 1, 2, \ldots N \qquad (7.10)$$

This first order equation can be solved by iteration, giving

$$F_k(s) = s^k F(s) - \sum_{j=0}^{k-1} s^j y^{(k-1-j)}(0) \qquad k = 0, 1 \ldots N \qquad (7.11)$$

from which

$$\int_0^\infty e^{-sx} Ly \, dx = F(s) \sum_{k=0}^{N} a_k s^k - \sum_{k=0}^{N} a_k \sum_{j=0}^{k-1} s^j y^{(k-1-j)}(0)$$

$$= F(s) \sum_{k=0}^{N} a_k s^k - \sum_{j=0}^{N-1} s^j \sum_{k=j}^{N-1} a_{k+1} y^{(k-j)}(0) = f(s) \qquad (7.12)$$

where

$$f(s) = \int_0^\infty e^{-sx} f(x) \, dx \qquad (7.13)$$

Thus

$$F(s) = \frac{\sum_{j=0}^{N-1} s^j \sum_{k=j}^{N-1} a_{k+1} y^{(k-j)}(0)}{\sum_{k=0}^{N} a_k s^k} + \frac{f(s)}{\sum_{k=0}^{N} a_k s^k}$$

$$= \frac{Q_{N-1}(s)}{P_N(s)} + \frac{f(s)}{P_N(s)} \qquad (7.14)$$

where $Q_{N-1}(s)$ and $P_N(s)$ are polynomials of order $N-1$ and N respectively.

Our goal is to obtain the solution $y(x)$ to the equation $Ly = f(x)$: This is given by the inverse Laplace transform

$$y(x) = \mathcal{L}^{-1}F(s) = \mathcal{L}^{-1}\left\{\frac{Q_{N-1}(s)}{P_N(s)}\right\} + \mathcal{L}^{-1}\left\{\frac{f(s)}{P_N(s)}\right\} \tag{7.15}$$

The solution to the homogeneous equation $Ly = 0$ is given by the first term on the right hand side of (7.15), in which the argument is the ratio of two polynomials in s, which can, in principal, be written as a sum of partial fractions, for the which the inverse Laplace transform is

$$\mathcal{L}^{-1}\left\{\frac{1}{(s-a)^{m+1}}\right\} = \frac{x^m e^{ax}}{m!} \tag{7.16}$$

Writing Q_{N-1}/P_N as a sum of partial fractions, however, requires that we determine the zeros of $P_N(s) = \sum_{k=0}^{N} a_k s^k$. This is straightforward for an equation of second degree ($N = 2$), possible with a bit more difficulty for an equation of third or fourth degree, but may not be possible analytically for $N \geq 5$.

The steps given thus far give the solution of the homogeneous equation $Ly = 0$ ($f(s) = 0$). For the inhomogeneous equation $Ly = f(x)$, if $f(x)$ is either a polynomial in x, exponential in form ($e^{\alpha x} \cdot \sin \alpha x$, $\cos \alpha x$) or a sum of products of polynomials and exponential functions, then the Laplace transform, $f(s)$, is a sum of terms of the form $(s-a)^{-m-1}$ (note Eq. (7.16)), and the second term in Eq. (7.14) can also be written as a sum of partial fractions, for which the inverse Laplace transform is again given by (7.16). In the case of a more general function $f(x)$, the inverse Laplace transform of $f(s)/P_N$ is given by the convolution theorem for Laplace transforms.

Given two functions, $f_1(x)$ and $f_2(x)$, with corresponding Laplace transforms $F_1(s)$ and $F_2(s)$,

$$\mathcal{L}f_1 = F_1(s) \qquad \mathcal{L}f_2 = F_2(s) \tag{7.17}$$

then

$$\mathcal{L}f_1\mathcal{L}f_2 = \mathcal{L}(f_1 * f_2) \tag{7.18}$$

where

$$f_1 * f_2 = \int_0^x f_1(t)f_2(x-t)\,dt = \int_0^x f_1(x-t)f_2(t)\,dt \tag{7.19}$$

is the convolution of f_1 and f_2.

We illustrate this explicitly by deriving the solution to the second degree linear inhomogeneous equation as given in (7.6) with $N = 2$ and $f(x)$ arbitrary. Setting

$$f_1(x) = f(x) \qquad F_1(s) = f(s)$$

$$f_2(x) = \mathcal{L}^{-1}\left\{\frac{1}{P_2(s)}\right\} \qquad F_2(s) = \frac{1}{P_2(s)} \tag{7.20}$$

we use the convolution theorem to obtain the inverse Laplace transform of the second term in (7.14):

$$\mathcal{L}^{-1}\left\{\frac{f(s)}{P_2(s)}\right\} = \mathcal{L}^{-1}\{\mathcal{L}f_1\mathcal{L}f_2\} = f_1 * f_2 \tag{7.21}$$

Here

$$P_2(s) = a_2 s^2 = a_1 s + a_0 = a_2(s - \alpha_1)(s - \alpha_2) \tag{7.22}$$

where

$$\alpha_1 = \frac{-a_1 + \Delta}{2a_2}, \quad \alpha_2 = \frac{-a_1 - \Delta}{2a_2}, \quad \Delta = \sqrt{a_1^2 - 4a_0 a_2} \tag{7.23}$$

from which

$$\frac{1}{P_2(s)} = \frac{1}{a_2(\alpha_1 - \alpha_2)}\left[\frac{1}{s - \alpha_1} - \frac{1}{s - \alpha_2}\right] \tag{7.24}$$

and, from (7.16),

$$f_2(x) = \mathcal{L}^{-1}\left\{\frac{1}{P_2(s)}\right\} = \frac{(e^{\alpha_1 x} - e^{\alpha_2 x})}{a_2(\alpha_1 - \alpha_2)} = \frac{2}{\Delta}e^{-\frac{a_1}{2a_2}}\sinh\left(\frac{\Delta x}{2a_2}\right) \tag{7.25}$$

Thus, from (7.21) we have

$$\mathcal{L}^{-1}\left\{\frac{f(s)}{P_2(s)}\right\} = \frac{2}{\Delta}\int_0^x f(x - t)e^{-\frac{a_1 t}{2a_2}}\sinh\left(\frac{\Delta t}{2a_2}\right)dt \tag{7.26}$$

Here, in the limit $\alpha_1 \to \alpha_2$ $(\Delta \to 0)$, we have

$$\mathcal{L}^{-1}\left\{\frac{f(s)}{P_2(s)}\right\} = \frac{1}{a_2}\int_0^x tf(x - t)\,dt = \frac{1}{a_2}\int_0^x (x - t)f(t)\,dt \tag{7.27}$$

It may be noted that we can also write (7.26) in the form

$$\mathcal{L}^{-1}\left\{\frac{f(s)}{P_2(s)}\right\} = \frac{1}{\Delta}\left[e^{\alpha_1 x}\int_0^x f(t)e^{-\alpha_1 t} - e^{\alpha_2 x}\int_0^x f(t)e^{-\alpha_2 t}\right], \tag{7.28}$$

which expresses $\mathcal{L}^{-1}\left\{\frac{f(s)}{P_2(s)}\right\}$ in terms of the finite Laplace transform of $f(x)$.

Finally, we evaluate the inverse Laplace transform of the first term in (7.14): From (7.16) with $N = 2$, we have

$$\mathcal{L}^{-1}\left\{\frac{Q_1(s)}{P_2(s)}\right\} = \frac{a_1 y(0) + a_2 y^{(1)}(0)}{a_2(\alpha_1 - \alpha_2)}\left[e^{\alpha_1 x} - e^{\alpha_2 x}\right] + \frac{y(0)}{(\alpha_1 - \alpha_2)}\left[\alpha_1 e^{\alpha_1 x} - \alpha_2 e^{\alpha_2 x}\right] \tag{7.29}$$

Writing this expression so that the limit $\alpha_1 \to \alpha_2$ $(\Delta \to 0)$ is evident, we have

$$\mathcal{L}^{-1}\left\{\frac{Q_1(s)}{P_2(s)}\right\} = y(0)e^{-\frac{a_1 x}{2a_2}}\left[\frac{a_1}{\Delta}\sinh\left(\frac{\Delta x}{2a_2}\right) + \cosh\left(\frac{\Delta x}{2a_2}\right)\right] + y^{(1)}(0)e^{-\frac{a_1 x}{2a_2}}\left[\frac{2a_2}{\Delta}\sinh\left(\frac{\Delta x}{2a_2}\right)\right]$$

(7.30)

Thus, in summary, the solution to the second order inhomogeneous equation with constant coefficients,

$$a_2 y''(x) + a_1 y'(x) + a_0 y(x) = f(x),$$

(7.31)

is

$$y(x) = y(0)e^{-\frac{a_1 x}{2a_2}}\left[\frac{a_1}{\Delta}\sinh\left(\frac{\Delta x}{2a_2}\right) + \cosh\left(\frac{\Delta x}{2a_2}\right)\right] + y^{(1)}(0)e^{-\frac{a_1 x}{2a_2}}\left[\frac{2a_2}{\Delta}\sinh\left(\frac{\Delta x}{2a_2}\right)\right]$$
$$+ \frac{2}{\Delta}\int_0^x f(x-t)e^{-\frac{a_1 t}{2a_2}}\sinh\left(\frac{\Delta t}{2a_2}\right)dt$$

(7.32)

We note that the use of the Laplace transform for the solution of a linear homogeneous or inhomogeneous differential equation, $Ly = f(x)$, involves three essential points:

(1) If the coefficients in the differential equation are constants, then the transform of Ly is a rational function of the transform variable, which can then be written as a sum of partial fractions.
(2) The inverse transform of a partial fraction, $1/(s-a)^{m+1}$, can be expressed analytically and has a simple closed form, $x^m e^{ax}/m!$.
(3) The transform of the solution $y(x)$ has, from the inhomogeneous term $f(x)$, a term of the form $f(s)/P_N(s)$, where $P_N(s)$ is a polynomial in the transform variable s. The inverse transform of $f(s)/P_N(s)$ can be expressed as an integral over $f(x)$ using the convolution theorem.

It will be seen in the following application of the generating function that these three points are equally essential for the solution of a linear difference equation, homogeneous or inhomogeneous.

7.2 Generating Functions and the Solution of Linear Difference Equations with Constant Coefficients

We consider now a linear inhomogeneous difference equation with constant coefficients:

$$Ly = p_N y(n+N) + p_{N-1} y(n+N-1) + \cdots + p_0 y(n) = q(n)$$

(7.33)

As with the differential equation, the solution to this equation will be obtained from its generating function, defined by

$$G(\omega) = \mathcal{G}y(n) = \sum_{n=0}^{\infty} y(n)\omega^n \tag{7.34}$$

from which $y(n)$ is given by the inverse of $\mathcal{G}$:

$$y(n) = \mathcal{G}^{-1}G(\omega) \tag{7.35}$$

Using the difference equation $Ly = q(n)$, we first obtain the generating function of Ly and then express the generating function of $y(n)$ in terms of the generating function of Ly.

To obtain the generating function of Ly we define

$$G_k(\omega) = \sum_{n=0}^{\infty} y(n+k)\omega^n \qquad k = 0, 1, \ldots N \qquad (G_0(\omega) = G(\omega)) \tag{7.36}$$

from which

$$
\begin{aligned}
G_{k-1}(\omega) &= \sum_{n=0}^{\infty} y(n+k-1)\omega^n \\
&= \omega \sum_{n=0}^{\infty} y(n+k-1)\omega^{n-1} \\
&= \sum_{n=0}^{\infty} y(n+k)\omega^n + y(k-1) \\
&= \omega G_k(\omega) + y(k-1), \qquad k = 1, 2, \ldots N
\end{aligned}
\tag{7.37}
$$

Thus

$$G_k(\omega) = \frac{G_{k-1}(\omega) - y(k-1)}{\omega}, \qquad k = 1, 2, \ldots N \tag{7.38}$$

This first order equation can be solved by iteration, or directly from the definition of $G_k(\omega)$:

$$
\begin{aligned}
G_k(\omega) &= \sum_{n=0}^{\infty} y(n+k)\omega^n = \frac{1}{\omega^k} \sum_{n=0}^{\infty} y(n+k)\omega^{n+k} = \frac{1}{\omega^k} \left[\sum_{n=0}^{\infty} y(n)\omega^n - \sum_{n=0}^{k-1} y(n)\omega^n \right] \\
&= \frac{1}{\omega^k} \left[G(\omega) - \sum_{n=0}^{k-1} y(n)\omega^n \right]
\end{aligned}
\tag{7.39}
$$

We then have the generating function of Ly:

$$\mathcal{G}\{Ly\} = \sum_{n=0}^{\infty} Ly\,\omega^n = \sum_{n=0}^{\infty} \sum_{k=0}^{N} p_k\, y(n+k)\omega^n$$

$$= \sum_{k=0}^{N} p_k G_k(\omega) = \sum_{n=0}^{\infty} q(n)\omega^n = Q(\omega) \tag{7.40}$$

Substituting (7.39) in this equation we have

$$\sum_{k=0}^{N} \frac{p_k}{\omega^k} \left[G(\omega) - \sum_{n=0}^{k-1} y(n)\omega^n \right] = Q(\omega) \tag{7.41}$$

from which

$$G(\omega) = \frac{Q(\omega)}{\sum_{k=0}^{N} \frac{p_k}{\omega^k}} + \frac{\sum_{k=0}^{N} \frac{p_k}{\omega^k} \sum_{n=0}^{k-1} y(n)\omega^n}{\sum_{k=0}^{N} \frac{p_k}{\omega^k}} \tag{7.42}$$

We now write $G(\omega)$ so that it has a form similar to that given in (7.14) for the Laplace transform $F(s)$. To that end we write each of the sums in (7.42) as rational functions of the transform variable ω. For the sum in the denominator we have

$$\sum_{k=0}^{N} \frac{p_k}{\omega^k} = \frac{1}{\omega^N} \sum_{k=0}^{N} p_k \omega^{N-k} = \frac{1}{\omega^N} \sum_{k=0}^{N} p_{N-k}\omega^k \tag{7.43}$$

For the first term in (7.42) we then have, from (7.43),

$$\frac{Q(\omega)}{\sum_{k=0}^{N} \frac{p_k}{\omega^k}} = \frac{\omega^N Q(\omega)}{\sum_{k=0}^{N} p_{N-k}\omega^k} = \frac{1}{p_0} \left[Q(\omega) - \frac{\sum_{k=0}^{N-1} p_{N-k}\omega^k}{\sum_{k=0}^{N} p_{N-k}\omega^k} Q(\omega) \right]$$

$$= \frac{1}{p_0} Q(\omega) + \frac{T_{N-1}(\omega)}{S_N(\omega)} Q(\omega) \tag{7.44}$$

where $T_{N-1}(\omega)$ and $S_N(\omega)$ are polynomials in the transform variable ω, of order $N-1$ and N respectively:

$$T_{N-1}(\omega) = -\frac{1}{p_0} \sum_{k=0}^{N-1} p_{N-k}\omega^k \tag{7.45}$$

and

$$S_N(\omega) = \sum_{k=0}^{N} p_{N-k}\omega^k \tag{7.46}$$

For the double sum in the numerator of the second term in (7.42) we have

$$\sum_{k=0}^{N} \frac{p_k}{\omega^k} \sum_{n=0}^{k-1} y(n)\omega^n = \sum_{k=1}^{N} p_k \sum_{n=0}^{k-1} y(n)\omega^{n-k} = \sum_{k=0}^{N-1} p_{k+1} \sum_{n=0}^{k} y(n)\omega^{n-k-1} \quad (7.47)$$

Inverting the order of summation gives

$$\sum_{k=0}^{N} \frac{p_k}{\omega^k} \sum_{n=0}^{k-1} y(n)\omega^n = \sum_{n=0}^{N-1} \sum_{k=n}^{N-1} p_{k-1} y(n)\omega^{n-k-1} \quad (7.48)$$

and making the substitution of summation variable $j = k - n + 1$ we have

$$\sum_{k=0}^{N} \frac{p_k}{\omega^k} \sum_{n=0}^{k-1} y(n)\omega^n = \sum_{n=0}^{N-1} \sum_{j=1}^{N-n} p_{n+j} y(n)\omega^{-j} \quad (7.49)$$

Again inverting the order of summation gives

$$\sum_{k=0}^{N} \frac{p_k}{\omega^k} \sum_{n=0}^{k-1} y(n)\omega^n = \frac{1}{\omega^N} \sum_{j=1}^{N} \omega^{N-j} \sum_{n=0}^{N-j} p_{n+j} y(n) \quad (7.50)$$

and making the substitution of summation variable $k = N - j$ gives, finally,

$$\sum_{k=0}^{N} \frac{p_k}{\omega^k} \sum_{n=0}^{k-1} y(n)\omega^n = \frac{1}{\omega^N} \sum_{k=0}^{N-1} \omega^k \sum_{n=0}^{k} p_{N-k-n} y(n) \quad (7.51)$$

Substituting (7.43) and (7.51) in the second term on the right hand side of (7.42), we see that this term is indeed a rational function of the transform variable in which the numerator is a polynomial of order $N - 1$ and the denominator is a polynomial of order N:

$$\frac{\sum_{k=0}^{N} \frac{p_k}{\omega^k} \sum_{n=0}^{k-1} y(n)\omega^n}{\sum_{k=0}^{N} \frac{p_k}{\omega^k}} = \frac{\sum_{k=0}^{N-1} \omega^k \sum_{n=0}^{k} p_{N-k-n} y(n)}{\sum_{k=0}^{N} p_{N-k}\omega^k} = \frac{R_{N-1}(\omega)}{S_N(\omega)} \quad (7.52)$$

where

$$R_{N-1}(\omega) = \sum_{k=0}^{N-1} \omega^k \sum_{n=0}^{k} p_{N-k-n} y(n) \quad (7.53)$$

and $S_N(\omega)$ is defined in (7.46).

Thus, substituting (7.44) and (7.52) in (7.42), we have

$$G(\omega) = \frac{R_{N-1}(\omega)}{S_N(\omega)} + \frac{T_{N-1}(\omega)}{S_N(\omega)} Q(\omega) + \frac{1}{p_0} Q(\omega) \tag{7.54}$$

The solution, $y(n)$, to the equation $Ly = q(n)$ is given by the inverse of $\mathcal{G}$. From (7.54) and (7.44) we have

$$y(n) = \mathcal{G}^{-1} G(\omega) = \mathcal{G}^{-1} \left\{ \frac{R_{N-1}(\omega)}{S_N(\omega)} \right\} + \mathcal{G}^{-1} \left\{ \frac{T_{N-1}(\omega)}{S_N(\omega)} Q(\omega) \right\} + \frac{1}{p_0} \mathcal{G}^{-1} Q(\omega) \tag{7.55}$$

The solution to the homogeneous equation $Ly = 0$ is given by the first term on the right hand side of (7.55), in which the argument is the ratio of two polynomials in ω and can be written as a sum of partial fractions, for which the inverse transform is (see Appendix G)

$$\mathcal{G}^{-1} \left\{ \frac{1}{(\omega - a)^{m+1}} \right\} = \frac{(-1)^{m+1}}{a^{n+m+1}} \binom{n+m}{m} \tag{7.56}$$

For the inhomogeneous equation $Ly = q(n)$, if $q(n)$ is either a polynomial in n, exponential in form (a^n) or a sum of products of polynomials and exponentials, then the generating function $Q(\omega)$ can be written as a sum of terms of the form $(\omega - a)^{-m-1}$ and the term $\frac{T_{N-1}(\omega)}{S_N(\omega)} Q(\omega)$ can also be written as a sum of partial fractions, for which the inverse transform is again given by (7.56). The inverse transform of the term $\frac{1}{p_0} Q(\omega)$ in (7.55) is straightforward: $\frac{1}{p_0} \mathcal{G}^{-1} Q(\omega) = \frac{1}{p_0} q(n)$. In the case of a more general function $q(n)$, the inverse transform $\mathcal{G}^{-1} G(\omega)$ is given by the convolution theorem for generating functions (see Appendix G).

Given two functions $y_1(n)$ and $y_2(n)$, with corresponding generating functions $G_1(\omega)$ and $G_2(\omega)$,

$$\mathcal{G} y_1 = G_1(\omega) \qquad \mathcal{G} y_2 = G_2(\omega) \tag{7.57}$$

then

$$\mathcal{G} y_1 \mathcal{G} y_2 = \mathcal{G}(y_1 * y_2) \tag{7.58}$$

where

$$y_1 * y_2 = \sum_{k=0}^{n} y_1(k) y_2(n-k) = \sum_{k=0}^{n} y_1(n-k) y_2(k) \tag{7.59}$$

is the convolution of y_1 and y_2.

We illustrate this explicitly by deriving the solution to the second degree linear inhomogeneous difference equation as given in (7.33) with $N = 2$ and $q(n)$ arbitrary. Referring to (7.55), we set

$$y_1(n) = q(n) \qquad G_1(\omega) = Q(\omega).$$

$$y_2(n) = \mathcal{G}^{-1}\left\{ \frac{T_1(\omega)}{S_2(\omega)} \right\} \qquad G_2(\omega) = \frac{T_1(\omega)}{S_2(\omega)} = -\frac{p_2 + p_1\omega}{p_0[p_2 + p_1\omega + p_0\omega^2]} \qquad (7.60)$$

and use the convolution theorem to obtain the second term on the right hand side of (7.55):

$$\mathcal{G}^{-1}\left\{ \frac{T_1(\omega)}{S_2(\omega)} Q(\omega) \right\} = \mathcal{G}^{-1}\{\mathcal{G}y_1 \mathcal{G}y_2\} = y_1 * y_2 \qquad (7.61)$$

Here

$$p_2 + p_1\omega + p_0\omega^2 = p_0(w - \beta_1)(w - \beta_2) \qquad (7.62)$$

where

$$\beta_1 = \frac{-p_1 + \Delta}{2p_0}, \qquad \beta_1 = \frac{-p_1 - \Delta}{2p_0}, \qquad \Delta = \sqrt{p_1^2 - 4p_0 p_2} \qquad (7.63)$$

from which

$$\frac{T_1(\omega)}{S_2(\omega)} = -\frac{1}{p_0^2(\beta_1 - \beta_2)}\left[\frac{p_2 + \beta_1 p_1}{w - \beta_1} - \frac{p_2 + \beta_2 p_1}{w - \beta_2} \right]$$

$$\frac{R_1(\omega)}{S_2(\omega)} = \frac{1}{p_0(\beta_1 - \beta_2)}\left\{ \frac{[(p_2 + \beta_1 p_1)y(0) + \beta_1 p_0 y(1)]}{w - \beta_1} - \frac{[(p_2 + \beta_2 p_1)y(0) + \beta_2 p_0 y(1)]}{w - \beta_2} \right\} \qquad (7.64)$$

From (7.56) we now have

$$y_2(n) = \mathcal{G}^{-1}\left\{ \frac{T_1(\omega)}{S_2(\omega)} \right\} = \frac{1}{p_0^2(\beta_1 - \beta_2)}\left[\frac{p_2 + \beta_1 p_1}{\beta_1^{n+1}} - \frac{p_2 + \beta_2 p_1}{\beta_2^{n+1}} \right] \qquad (7.65)$$

and from (7.61) to (7.59) we have

$$\mathcal{G}^{-1}\left\{ \frac{T_1(\omega)}{S_2(\omega)} Q(\omega) \right\} = y_1 * y_2 = \frac{1}{p_0^2(\beta_1 - \beta_2)}\left\{ \frac{p_2 + \beta_1 p_1}{\beta_1^{n+1}} \sum_{k=0}^{n} q(k)\beta_1^k - \frac{p_2 + \beta_2 p_1}{\beta_2^{n+1}} \sum_{k=0}^{n} q(k)\beta_2^k \right\} \qquad (7.66)$$

We note that, similar to the analysis for the differential equation, the inverse transform is expressed in terms of the finite generating function of $q(n)$.

Finally, we evaluate the inverse transform of the first term on the right hand side of (7.55). From (7.64) to (7.56) we have

$$\mathcal{G}^{-1}\left\{ \frac{R_1(\omega)}{S_2(\omega)} \right\} = -\frac{1}{p_0(\beta_1 - \beta_2)}\left\{ \frac{[(p_2 + \beta_1 p_1)y(0) + \beta_1 p_0 y(1)]}{\beta_1^{n+1}} - \frac{[(p_2 + \beta_2 p_1)y(0) + \beta_2 p_0 y(1)]}{\beta_2^{n+1}} \right\} \qquad (7.67)$$

In summary, the solution to the second order inhomogeneous difference equation with constant coefficients,

$$p_2 y(n+2) + p_1 y(n+1) + p_0 y(n) = q(n), \tag{7.68}$$

is

$$
y(n) = -\frac{1}{p_0(\beta_1 - \beta_2)} \left\{ \frac{[(p_2 + \beta_1 p_1) y(0) + \beta_1 p_0 y(1)]}{\beta_1^{n+1}} - \frac{[(p_2 + \beta_2 p_1) y(0) + \beta_2 p_0 y(1)]}{\beta_2^{n+1}} \right\}
$$
$$
+ \frac{1}{p_0^2(\beta_1 - \beta_2)} \left\{ \frac{p_2 + \beta_1 p_1}{\beta_1^{n+1}} \sum_{k=0}^{n} q(k)\beta_1^k - \frac{p_2 + \beta_2 p_1}{\beta_2^{n+1}} \sum_{k=0}^{n} q(k)\beta_2^k \right\} + \frac{1}{p_0} q(n) \tag{7.69}
$$

7.3   Laplace Transforms and the Solution of Linear Differential Equations with Polynomial Coefficients

We now consider the extension of the analysis given thus far to the case of differential and difference equations in which the coefficients are polynomials, in x for differential equations, in n for difference equations. We begin with the differential equation

$$Ly = a_N(x) y^{(N)}(x) + a_{N-1}(x) y^{(N-1)}(x) + \cdots + a_0(x) y(x) = f(x), \tag{7.70}$$

in which $a_k(x)$, $k = 0, 1, \ldots N$, is a polynomial of degree M_k:

$$a_k(x) = \sum_{m=0}^{M_k} a_{km} x^m \tag{7.71}$$

and the a_{km} are constants, but

$$a_{km} \neq 0 \quad \text{for } m = M_k \tag{7.72}$$

Thus

$$Ly = \sum_{k=0}^{N} a_k(x) y^{(k)}(x) = \sum_{k=0}^{N} \sum_{m=0}^{M_k} a_{km} x^m y^{(k)}(x) \tag{7.73}$$

As in (7.9), we define

$$F_k(s) = \int_0^\infty e^{-sx} y^{(k)}(x)\, dx, \qquad k = 0, 1, \ldots N, \qquad (F_0(s) = F(s)) \tag{7.74}$$

from which, from (7.9)–(7.11),

$$F_k(s) = s^k F(s) - \sum_{j=0}^{k-1} s^j y^{(k-1-j)}(0), \qquad k = 0, 1 \ldots N \qquad (7.75)$$

We then have

$$
\begin{aligned}
\int_0^\infty e^{-sx} a_k(x) y^{(k)}(x)\, dx &= \sum_{m=0}^{M_k} a_{km} \int_0^\infty e^{-sx} x^m y^{(k)}\, dx \\
&= \sum_{m=0}^{M_k} a_{km} (-1)^m \frac{d^m}{ds^m} \int_0^\infty e^{-sx} y^{(k)}(x)\, dx \\
&= \sum_{m=0}^{M_k} (-1)^m a_{km} \frac{d^m}{ds^m} F_k(s) \\
&= \sum_{m=0}^{M_k} (-1)^m a_{km} \frac{d^m}{ds^m} \left[s^k F(s) - \sum_{j=0}^{k-1} s^j y^{(k-1-j)}(0) \right]
\end{aligned}
$$

$$(7.76)$$

The Laplace transform of the differential equation (7.70) is then

$$
\begin{aligned}
\mathcal{L}(Ly) &= \int_0^\infty e^{-sx} Ly\, dx \\
&= \sum_{k=0}^{N} \sum_{m=0}^{M_k} (-1)^m a_{km} \frac{d^m}{ds^m} \left[s^k F(s) - \sum_{j=0}^{k-1} s^j y^{(k-1-j)}(0) \right] = f(s) \quad (7.77)
\end{aligned}
$$

where

$$f(s) = \int_0^\infty e^{-sx} f(x)\, dx \qquad (7.78)$$

In the summation over m in (7.77) we consider the term $m = M_k$:

$$\sum_{k=0}^{N} (-1)^{M_k} a_{kM_k} \frac{d^{M_k}}{ds^{M_k}} \left[s^k F(s) - \sum_{j=0}^{k-1} s^j y^{(k-1-j)}(0) \right] \qquad (7.79)$$

The highest derivative of $F(s)$ in this expression is

$$\sum_{k=0}^{N} (-1)^{M_k} a_{kM_k} s^k F^{(M_k)}(s) \qquad (7.80)$$

Equation (7.77) is thus a differential equation for the Laplace transform, $F(s)$, whose order is equal to the highest degree of the polynomial coefficients $a_k(x)$: $\max(M_k)$, $k = 0, 1, \ldots, N$. In this differential equation, the derivatives of $F(s)$ have coefficients which are polynomials in the transform variable, s; the degree of the coefficient of highest degree is equal to N, the order of the original differential equation in x, (7.70). It is clear that this is not a reasonable method for the solution of the differential equation $Ly = f(x)$ given in (7.70) unless the coefficients $a_k(x)$ are either linear or quadratic functions of x.

For the linear differential equation with linear coefficients we may write Eqs. (7.70) and (7.71) in the form

$$Ly = \sum_{k=0}^{N} p_k(x) y^{(k)}(x) = f(x) \tag{7.81}$$

in which

$$p_k(x) = c_k + d_k x \tag{7.82}$$

We then have

$$\mathcal{L}(Ly) = \sum_{k=0}^{N} \int_0^\infty e^{-sx} p_k(x) y^{(k)}(x)\, dx$$

$$= \sum_{k=0}^{N} c_k \int_0^\infty e^{-sx} y^{(k)}(x)\, dx + \sum_{k=0}^{N} d_k \int_0^\infty e^{-sx} x y^{(k)}(x)\, dx$$

$$= \sum_{k=0}^{N} c_k F_k(s) - \sum_{k=0}^{N} d_k F_k'(s) \tag{7.83}$$

where, from (7.9) to (7.11),

$$F_k(s) = s^k F(s) - \sum_{j=0}^{k-1} s^j y^{(k-1-j)}(0), \qquad k = 0, 1 \ldots N, \qquad (F_0 = F(s)) \tag{7.84}$$

Then, with

$$f_0(s) = \sum_{k=0}^{N} c_k s^k$$

$$\tag{7.85}$$

$$f_1(s) = \sum_{k=0}^{N} d_k s^k$$

we can write the Laplace transform of the nth order inhomogeneous differential equation, $\mathcal{L}(Ly) = f(s)$, as a first-order differential equation for the Laplace transform, $F(s)$, of its solution:

$$f_0 F(s) - (f_1(s)F(s))' = g(s) + f(s) \tag{7.86}$$

where

$$g(s) = \sum_{k=1}^{N} \left(c_k g_k(s) - d_k g_k'(s) \right) \tag{7.87}$$

in which

$$g_k(s) = \sum_{j=0}^{k-1} s^j y^{k-1-j}(0) \tag{7.88}$$

(Note (1.14)) which may be written as

$$\left(f_1(s)e^{-\int \frac{f_0(s)}{f_1(s)} ds} F(s) \right)' = -e^{-\int \frac{f_0(s)}{f_1(s)} ds} (g(s) + f(s)) \tag{7.89}$$

from which

$$F(s) = -f_1^{-1}(s)e^{\int \frac{f_0(s)}{f_1(s)} ds} \int e^{-\int \frac{f_0(s)}{f_1(s)} ds} (g(s) + f(s)) \, ds. \tag{7.90}$$

However, even in the case of a second-order homogeneous differential equation with linear coefficients, obtaining the solution explicitly by inverting the Laplace transform is not without difficulty. Thus, from (7.90), with $N = 2$ and $f(s) = 0$ we have

$$F(s) = f_1^{-1}(s)e^{\int \frac{f_0(s)}{f_1(s)} ds} \int (a + bs)e^{-\int \frac{f_0(s)}{f_1(s)} ds} \, ds \tag{7.91}$$

in which

$$a = -c_1 y(0) - c_2 y'(0) + d_2 y(0)$$
$$b = -c_2 y(0) \tag{7.92}$$

If for now we assume that $d_2 \neq 0$ and $d_1^2 - 4d_0 d_2 \neq 0$ so that the roots of $f_1(s) = 0$ are distinct, then from (7.85), (with $N = 2$),

$$f_1(s) = d_2(s - \alpha_1)(s - \alpha_2) \tag{7.93}$$

in which

$$\alpha_1 = \frac{-d_1 + \Delta}{2d_2}, \qquad \alpha_2 = \frac{-d_1 - \Delta}{2d_2} \tag{7.94}$$

and

$$d_2(\alpha_1 - \alpha_2) = \sqrt{d_1^2 - 4d_0 d_2} \equiv \Delta \tag{7.95}$$

from which

$$\frac{f_0(s)}{f_1(s)} = \frac{c_2}{d_2} + \frac{\gamma_1}{(s - \alpha_1)} + \frac{\gamma_2}{(s - \alpha_2)} \tag{7.96}$$

where

$$\gamma_1 = \frac{f_0(\alpha_1)}{\Delta}$$

$$\gamma_2 = -\frac{f_0(\alpha_2)}{\Delta} \tag{7.97}$$

and

$$e^{-\int \frac{f_0(s)}{f_1(s)} ds} = e^{-\frac{c_2}{d_2} s}(s - \alpha_1)^{-\gamma_1}(s - \alpha_2)^{-\gamma_2} \tag{7.98}$$

(We note the relation of the exponents γ_1 and γ_2 to the exponents β_1 and β_2 for the second order difference equation given in (7.161).)

We may then substitute this in (7.91) and obtain an integral for the Laplace transform $F(s)$. The solution $y(x)$ to the differential equation is then given by the inverse Laplace transform:

$$y(x) = \mathcal{L}^{-1} F(s) = \frac{1}{2\pi i} \int_{\gamma - i\infty}^{\gamma + i\infty} e^{sx} F(s) \, ds \tag{7.99}$$

In principle this analysis gives the solution $y(x)$ with specified initial values, $y(0)$ and $y'(0)$. It is clear, however, that obtaining an explicit solution in terms of the classical functions of mathematical physics following this approach is not a simple matter. We therefore present, in the following section, an alternate approach which is often taken. This consists in writing the solution to the differential equation as an integral similar to the inverse Laplace transform given above in (7.99) except that the limits of integration as well as the transform in the integrand are determined by using the equation $Ly = f(x)$ to derive the necessary conditions that the transform must satisfy. (See, e.g., [25, Chap. 5, Sect. 47] and [27, A, Sects. 5, 19]). We consider in detail the second order homogeneous equation with linear coefficients with the view of comparing the solution with that of the corresponding difference equation.

7.4   Alternative Method for the Solution of Homogeneous Linear Differential Equations with Linear Coefficients

We consider here the homogenous differential equation

$$\sum_{k=0}^{N} p_k(x) y^{(k)}(x) = 0 \tag{7.100}$$

in which the coefficients $p_k(x)$ are linear functions of x:

$$p_k(x) = c_k + d_k x \tag{7.101}$$

In this alternative method, a solution of (7.100) is sought in the form of a Laplace transform:

$$y(x) = \int e^{-sx} v(s) \, ds \tag{7.102}$$

in which, as mentioned, both $v(s)$ and the integration path are to be determined after using the equation $Ly = f(x)$ to derive the necessary conditions that the transform must satisfy. (The relation to the analysis just given using the Laplace transform and its inverse is seen if we write $s = -t$ and $v(s) = v(-t) = u(t)$.)

Then

$$y^{(k)}(x) = (-1)^k \int e^{-sx} s^k v(s) \, ds$$

$$xy^{(k)}(x) = (-1)^k \int s^k v(s) \, d(-e^{-sx})$$

$$= -(-1)^k e^{-sx} s^k v(s) \Big| + (-1)^k \int e^{-sx} (s^k v(s))' \, ds \tag{7.103}$$

Defining

$$g_0(s) = f_0(-s) = \sum_{k=0}^{N} (-1)^k c_k s^k$$

$$\tag{7.104}$$

$$g_1(s) = f_1(-s) = \sum_{k=0}^{N} (-1)^k d_k s^k$$

(Here c_N and d_N are arbitrary except that they can not both be zero.)

We then have

$$Ly = \sum_{k=0}^{N} p_k(x) y^{(k)}(x)$$

$$= -e^{-sx} g_1(s) v(s) \Big| + \int e^{-sx} [(g_0(s) + g_1'(s)) v(s) + g_1(s) v'(s)] \, ds \quad (7.105)$$

Now in order to satisfy (7.100) we choose $v(s)$ to satisfy the first order differential equation

$$(g_0(s) + g_1'(s)) v(s) + g_1(s) v'(s) = 0 \quad (7.106)$$

and choose the integration path such that

$$e^{-sx} g_1(s) v(s) \Big| = 0 \quad (7.107)$$

For an open path of integration, $e^{-sx} g_1(s) v(s)$ must have the same value at the two end points. For a closed path, $e^{-sx} g_1(s) v(s)$ must return to its initial value.

From (7.106) we have

$$\frac{(v(s) g_1(s))'}{v(s) g_1(s)} = -\frac{g_0(s)}{g_1(s)} \quad (7.108)$$

from which, on integrating,

$$v(s) = \frac{A}{g_1(s)} \exp\left(-\int \frac{g_0(s)}{g_1(s)} \, ds\right) \quad (7.109)$$

The integration in (7.109) is straightforward in that $g_0(s)/g_1(s)$ is the ratio of two polynomials and can be expressed as the sum of partial fractions and a polynomial in s.

If we now consider the second order homogeneous differential equation, then from (7.104) to (7.85)

$$\frac{g_0(s)}{g_1(s)} = \frac{c_2}{d_2} - \frac{\gamma_1}{(s + \alpha_1)} - \frac{\gamma_2}{(s + \alpha_2)} \quad (7.110)$$

where from (7.93)

$$g_1(s) = d_2(s + \alpha_1)(s + \alpha_2). \quad (7.111)$$

Then from (7.109)

$$v(s) = A e^{-\frac{c_2}{d_2} s} (s + \alpha_1)^{\gamma_1 - 1} (s + \alpha_2)^{\gamma_2 - 1} \quad (7.112)$$

Substituting $v(s)$ in (7.102), we have

$$.y(x) = A \int e^{-(x+\frac{c_2}{d_2})s}(s + \alpha_2)^{\gamma_1-1}(s + \alpha_2)^{\gamma_2-1} \, ds \qquad (7.113)$$

in which the integration path must be chosen such that

$$\left. e^{-sx}g_1(s)v(s) \right| = 0, \qquad (7.114)$$

the particular choice depending on the values of γ_1 and γ_2. With the change of integration variable $s = (\alpha_1 - \alpha_2)\sigma - \alpha_1$, we have

$$y(x) = Ae^{\alpha_1(x+\frac{c_2}{d_2})} \int e^{-(\alpha_1-\alpha_2)(x+\frac{c_2}{d_2})\sigma}\sigma^{\gamma_1-1}(1-\sigma)^{\gamma_2-1} \, d\sigma \qquad (7.115)$$

Here, and throughout the rest of this section, the factor A is taken to be a generic factor that includes all factors which are independent of x.

The significant point to note, on substituting $v(s)$ as given above in (7.109) in (7.102), is that $y(x)$ is then expressed as an integral in which the integrand has the factor e^{-sx}, which is characteristic of the confluent hypergeometric function.

If $\Re\gamma_1 > 0$ and $\Re\gamma_2 > 0$ then we can choose $\sigma = 0$ and $\sigma = 1$ as end points of the integration and (7.114) is satisfied. We then have a solution (see [13, Sect. 6.5(1)])

$$\begin{aligned}
y_1(x) &= Ae^{\alpha_1(x+\frac{c_2}{d_2})} \int_0^1 e^{-(\alpha_1-\alpha_2)(x+\frac{c_2}{d_2})\sigma}\sigma^{\gamma_1-1}(1-\sigma)^{\gamma_2-1} \, d\sigma \\
&= A\frac{\Gamma(\gamma_1)\Gamma(\gamma_2)}{\Gamma(\gamma_1+\gamma_2)}e^{\alpha_1(x+\frac{c_2}{d_2})}{}_1F_1\left(\gamma_1; \gamma_1+\gamma_2; -(\alpha_1-\alpha_2)\left(x+\frac{c_2}{d_2}\right)\right) \\
&= A\frac{\Gamma(\gamma_1)\Gamma(\gamma_2)}{\Gamma(\gamma_1+\gamma_2)}e^{\alpha_2(x+\frac{c_2}{d_2})}{}_1F_1\left(\gamma_2; \gamma_1+\gamma_2; (\alpha_1-\alpha_2)\left(x+\frac{c_2}{d_2}\right)\right) \quad (7.116)
\end{aligned}$$

A second linearly independent solution of the Eq. (7.100) with $N = 2$, provided that $\gamma_1 + \gamma_2$ is not an integer, is (see [13, Sect. 6.3(3)])

$$\begin{aligned}
y_2(x) &= A\left(x+\frac{c_2}{d_2}\right)^{1-\gamma_1-\gamma_2} e^{\alpha_1(x+\frac{c_2}{d_2})}{}_1F_1\left(1-\gamma_2; 2-\gamma_1-\gamma_2; -(\alpha_1-\alpha_2)\left(x+\frac{c_2}{d_2}\right)\right) \\
&= A\left(x+\frac{c_2}{d_2}\right)^{1-\gamma_1-\gamma_2} e^{\alpha_2(x+\frac{c_2}{d_2})}{}_1F_1\left(1-\gamma_1; 2-\gamma_1-\gamma_2; (\alpha_1-\alpha_2)\left(x+\frac{c_2}{d_2}\right)\right)
\end{aligned}$$
$$(7.117)$$

The general solution of (7.100) with $N = 2$ is then

$$y(x) = Ay_1(x) + By_2(x) \qquad (7.118)$$

The case in which $\gamma_1 + \gamma_2$ is an integer is treated in detail in [13, Sect. 6.7.1].

If only one of the constants $\Re\gamma_1$ and $\Re\gamma_2$ is positive, then the integral (7.113) may be evaluated using a closed contour as given in [13, Sect. 6.11.1(2) and (3)].

If $\Re\gamma_1 > 0$, then we may write

$$
y_1(x) = Ae^{\alpha_1(x+\frac{c_2}{d_2})} \int_0^{(1+)} e^{-(\alpha_1-\alpha_2)(x+\frac{c_2}{d_2})\sigma} \sigma^{\gamma_1-1}(\sigma-1)^{\gamma_2-1}\,d\sigma
$$

$$
= A\frac{2\pi i\,\Gamma(\gamma_1)}{\Gamma(\gamma_1+\gamma_2)\Gamma(1-\gamma_2)} e^{\alpha_1(x+\frac{c_2}{d_2})}\,_1F_1\left(\gamma_1;\gamma_1+\gamma_2;-(\alpha_1-\alpha_2)\left(x+\frac{c_2}{d_2}\right)\right)
$$

$$
= A\frac{2\pi i\,\Gamma(\gamma_1)}{\Gamma(\gamma_1+\gamma_2)\Gamma(1-\gamma_2)} e^{\alpha_2(x+\frac{c_2}{d_2})}\,_1F_1\left(\gamma_2;\gamma_1+\gamma_2;(\alpha_1-\alpha_2)\left(x+\frac{c_2}{d_2}\right)\right)
$$

$$
(7.119)
$$

If $\Re\gamma_2 > 0$, then we may write

$$
y_1(x) = Ae^{\alpha_1(x+\frac{c_2}{d_2})} \int_1^{(0+)} e^{-(\alpha_1-\alpha_2)(x+\frac{c_2}{d_2})\sigma} (-\sigma)^{\gamma_1-1}(1-\sigma)^{\gamma_2-1}\,d\sigma
$$

$$
= A\frac{2\pi i\,\Gamma(\gamma_2)}{\Gamma(\gamma_1+\gamma_2)\Gamma(1-\gamma_1)} e^{\alpha_1(x+\frac{c_2}{d_2})}\,_1F_1\left(\gamma_1;\gamma_1+\gamma_2;-(\alpha_1-\alpha_2)\left(x+\frac{c_2}{d_2}\right)\right)
$$

$$
= A\frac{2\pi i\,\Gamma(\gamma_2)}{\Gamma(\gamma_1+\gamma_2)\Gamma(1-\gamma_1)} e^{\alpha_2(x+\frac{c_2}{d_2})}\,_1F_1\left(\gamma_2;\gamma_1+\gamma_2;(\alpha_1-\alpha_2)\left(x+\frac{c_2}{d_2}\right)\right)
$$

$$
(7.120)
$$

If both $\Re\gamma_1 < 0$ and $\Re\gamma_2 < 0$ then the integral may be evaluated using the closed loop contour given in [13, Sect. 6.11.1(1)] and we may then write

$$
y_1(x) = Ae^{\alpha_1(x+\frac{c_2}{d_2})} \int^{(1+,0+,1-,0-)} e^{-(\alpha_1-\alpha_2)(x+\frac{c_2}{d_2})\sigma} \sigma^{\gamma_1-1}(1-\sigma)^{\gamma_2-1}\,d\sigma
$$

$$
= A\frac{(2\pi i)^2 e^{i\pi(\gamma_1+\gamma_2)}}{\Gamma(1-\gamma_1)\Gamma(1-\gamma_2)\Gamma(\gamma_1+\gamma_2)} e^{\alpha_1(x+\frac{c_2}{d_2})}\,_1F_1\left(\gamma_1;\gamma_1+\gamma_2;-(\alpha_1-\alpha_2)\left(x+\frac{c_2}{d_2}\right)\right)
$$

$$
= A\frac{(2\pi i)^2 e^{i\pi(\gamma_1+\gamma_2)}}{\Gamma(1-\gamma_1)\Gamma(1-\gamma_2)\Gamma(\gamma_1+\gamma_2)} e^{\alpha_2(x+\frac{c_2}{d_2})}\,_1F_1\left(\gamma_2;\gamma_1+\gamma_2;(\alpha_1-\alpha_2)\left(x+\frac{c_2}{d_2}\right)\right)
$$

$$
(7.121)
$$

In all these cases the term in (7.114) is zero, and the solution $y_1(x)$ is given by the confluent hypergeometric functions given above.

As will be seen in the next section, this provides both a similarity and a distinction when comparing the second order differential equation with the second order difference equation with linear coefficients. In the case of the second order homogeneous *difference* equation with linear coefficients, given below in (7.143), when any of the parameters d_k is zero, or when the roots of the equation $d_2 t^2 + d_1 t + d_0 = 0$ are equal, $v(t)$ as given in (7.152) again has an exponential factor, and the integral for $y(n)$, Eq. (7.137), can be expressed in terms of the confluent hypergeometric function. However, when the roots of this equation are distinct, and none of the parameters d_k is zero, then the solution of the difference equation leads to the well-known classical functions of mathematical physics such as the Legendre and Gegenbauer functions, and is expressed in terms of the hypergeometric function $_2F_1$.

7.5 Generating Functions and the Solution of Linear Difference Equations with Polynomial Coefficients

In this section we extend the analysis given thus far for difference equations with constant coefficients to the case of difference equations in which the coefficients are polynomials in n.[1] We begin with the difference equation

$$Ly = a_N(n)y(n+N) + a_{N-1}(n)y(n+N-1) + \cdots + a_0(n)y(n) = q(n), \quad (7.122)$$

in which $a_k(n)$, $k = 0, 1, \ldots, N$, is a polynomial of degree M_k:

$$a_k(n) = \sum_{m=0}^{M_k} a_{km}n^m \quad (7.123)$$

and the a_{km} are constants, but

$$a_{km} \neq 0 \quad \text{for } m = M_k \quad (7.124)$$

Thus

$$Ly = \sum_{k=0}^{N} a_k(n)y(n+k) = \sum_{k=0}^{N} \sum_{m=0}^{M_k} a_{km}n^m y(n+k) \quad (7.125)$$

From (7.39)

$$G_k(\omega) = \sum_{n=0}^{\infty} y(n+k)\omega^n$$

$$= \frac{1}{\omega^k} \left[G(\omega) - \sum_{n=0}^{k-1} y(n)\omega^n \right] \quad (7.126)$$

---

[1] See Weixlbaumer [42] for a mathematical analysis about the state of the art concerning the search for solutions of linear difference equations. Algorithms are presented for finding polynomial, rational, hypergeometric and d'Alembertian solutions.

Thus

$$\mathcal{G}(Ly) = \sum_{n=0}^{\infty} Ly\,\omega^n = \sum_{n=0}^{\infty}\sum_{k=0}^{N} a_k(n)y(n+k)\omega^n$$

$$= \sum_{n=0}^{\infty}\sum_{k=0}^{N}\sum_{m=0}^{M_k} a_{km}n^m y(n+k)\omega^n$$

$$= \sum_{n=0}^{\infty}\sum_{k=0}^{N}\sum_{m=0}^{M_k} a_{km}\left(\omega\frac{\partial}{\partial\omega}\right)^m y(n+k)\omega^n$$

$$= \sum_{k=0}^{N}\sum_{m=0}^{M_k} a_{km}\left(\omega\frac{\partial}{\partial\omega}\right)^m \sum_{n=0}^{\infty} y(n+k)\omega^n$$

$$= \sum_{k=0}^{N}\sum_{m=0}^{M_k} a_{km}\left(\omega\frac{\partial}{\partial\omega}\right)^m G_k(\omega)$$

$$= \sum_{k=0}^{N}\sum_{m=0}^{M_k} a_{km}\left(\omega\frac{\partial}{\partial\omega}\right)^m \left(\frac{1}{\omega^k}\left[G(\omega) - \sum_{n=0}^{k-1} y(n)\omega^n\right]\right)$$

$$\tag{7.127}$$

In (7.127) the term $m = M_k$ is

$$\sum_{k=0}^{N} a_{kM_k}\left(\omega\frac{\partial}{\partial\omega}\right)^{M_k}\left(\frac{1}{\omega^k}\left[G(\omega) - \sum_{n=0}^{k-1} y(n)\omega^n\right]\right) \tag{7.128}$$

The highest derivative of $G(\omega)$ is thus

$$\sum_{k=0}^{N} a_{kM_k}\omega^{M_k-k}G^{(M_k)}(\omega) \tag{7.129}$$

Thus, similar to the case of the differential equation with polynomial coefficients, we now obtain a differential equation for the generating function, $G(\omega)$, whose order is equal to the highest degree of the polynomial coefficients $a_k(n)$: $\max(M_k)$, $k = 0, 1, \ldots N$. Again it is clear that this is also not a reasonable method for the solution of difference equations unless the coefficients are either linear or quadratic functions of n. We therefore examine in some detail the case in which the coefficients are linear functions of n. This involves obtaining the solution, $y(n)$, by inversion of the generating function: $y(n) = \mathcal{G}^{-1}G(\omega)$, for which see Appendix G.

7.6 Solution of Homogeneous Linear Difference Equations with Linear Coefficients

We now consider the homogeneous difference equation

$$\sum_{k=0}^{N} p_k(n)y(n+k) = 0 \tag{7.130}$$

in which the coefficients $p_k(n)$ are linear functions of n:

$$p_k(n) = a_k + b_k n \tag{7.131}$$

Although our primary consideration is for the case in which n is an integer, we note that the analysis is essentially unchanged if we replace n by a non-integer variable x. We then have the difference equation

$$\sum_{k=0}^{N} p_k(x)y(x+k) = 0 \tag{7.132}$$

in which

$$p_k(x) = a_k + b_k x \tag{7.133}$$

(Purists characterize Eq. (7.130) as a recursion relation, and Eq. (7.132) as a difference equation).

We note that if $y(x)$ is a solution of the homogeneous equation (7.132), then a more general solution is given by $f(x)y(x)$, where $f(x)$ is a function satisfying

$$f(x+1) = f(x), \tag{7.134}$$

for example

$$f(x) = A + B\sin(2\pi(x+\beta)) + C\cos(2\pi(x+\gamma)) \tag{7.135}$$

or

$$f(x) = A + Be^{2\pi x i} \tag{7.136}$$

We seek a solution of Eq. (7.132) having the form of a Mellin transform:

$$y(x) = \int t^{x-1}v(t)\,dt \tag{7.137}$$

(Note that if we let $t = e^{-s}$ we have the usual form of the Laplace transform:

$$y(x) = \int e^{-sx} V(s)\, ds \tag{7.138}$$

in which $V(s) = -v(t)$.) In (7.137), both $v(t)$ and the integration path are to be determined. From (7.137) we have

$$y(x + k) = \int t^{x+k-1} v(t)\, dt \tag{7.139}$$

Substituting this in (7.132) we have

$$\sum_{k=0}^{N} p_k(x) \int t^{x+k-1} v(t)\, dt = 0 \tag{7.140}$$

This expression suggests that rather than defining $p_k(x)$ as in (7.133), we write

$$p_k(x) = c_k + d_k(x + k), \tag{7.141}$$

which constitutes a simple change of parameters:

$$d_k = b_k, \qquad c_k + k d_k = a_k. \tag{7.142}$$

Equation (7.132) then has the form

$$\sum_{k=0}^{N} [c_k + d_k(x + k)] y(x + k) = 0 \tag{7.143}$$

Writing the coefficients in the form $p_k(x) = c_k + d_k(x + k)$, we may write the difference equation in the form

$$\sum_{k=i}^{N+i} [c_k + d_k(x + k)] y(x + k) = 0, \qquad i = 0, \pm 1, \pm 2, \ldots \tag{7.144}$$

and note that c_k and d_k remain independent of i. (This will be applied when we consider the second order difference equation ($N = 2$), and choose $i = -1$.) Had we taken the coefficients to have the form given in (7.133), viz., $p_k(x) = a_k + b_k x$, and written the summation as in (7.144), then a_k would depend on i : $a_k \to a_k - i b_k$. More significantly, as will be seen when we consider the second order difference equation, the parameters c_k and d_k are intrinsic to the functional behavior of the solution, specifically in the form c_k/d_k, ($k = 0, 1, 2$) and $d_0 d_2/d_1^2$. From (7.140) and (7.141) we now have

$$\sum_{k=0}^{N} [c_k \int t^{x+k-1} v(t)\, dt + d_k \int (x+k) t^{x+k-1} v(t)\, dt] = 0 \tag{7.145}$$

Here in the last term in square brackets, we have, integrating by parts,

$$\int (x+k) t^{x+k-1} v(t)\, dt = \int v(t)\, d(t^{x+k}) = t^{x+k} v(t) \Big| - \int t^{x+k-1} t v'(t)\, dt \tag{7.146}$$

Defining

$$f_0(t) = \sum_{k=0}^{N} c_k t^k$$

$$f_1(t) = \sum_{k=0}^{N} d_k t^k \tag{7.147}$$

we have, from (7.140),

$$\sum_{k=0}^{N} p_k(x) \int t^{x+k-1} v(t)\, dt = f_1(t) t^x v(t) \Big| + \int t^{x-1} [f_0(t) v(t) - f_1(t) t v'(t)]\, dt = 0 \tag{7.148}$$

Now, in order to satisfy (7.140) we choose $v(t)$ to satisfy the first order differential equation

$$f_0(t) v(t) - f_1(t) t v'(t) = 0 \tag{7.149}$$

and choose an integration path such that

$$f_1(t) t^x v(t) \Big| = 0 \tag{7.150}$$

For an open path of integration, $f_1(t) t^x v(t)$ must have the same value at the two end points. For a closed path, $f_1(t) t^x v(t)$ must return to its initial value. From (7.147) and (7.149) we have

$$\frac{v'(t)}{v(t)} = \frac{f_0(t)}{t f_1(t)} = \frac{\sum_{k=0}^{N} c_k t^k}{t \sum_{k=0}^{N} d_k t^k} \tag{7.151}$$

from which, on integrating,

$$v(t) = A \exp \left(\int \frac{f_0(t)}{t f_1(t)}\, dt \right) \tag{7.152}$$

(Here, and throughout the rest of this section, the factor A is taken to be a generic factor that includes all factors which are independent of x.) In the integrand on the right hand side of (7.152) we have an Nth order polynomial in the numerator and a polynomial of order $N + 1$ in the denominator (assuming that $d_N \neq 0$). This fraction can then be expressed as a sum of partial fractions and integrated. Analysis with arbitrary N may be found in [35, Chap. 11, Sects. 2 and 6, pp. 323–324, 343–345].

In order to present explicit solutions to the difference equation (7.132), we restrict ourselves at this point to second order homogeneous difference equations ($N = 2$):

$$(c_2 + d_2(x + 2))y(x + 2) + (c_1 + d_1(x + 1))y(x + 1) + (c_0 + d_0 x)y(x) = 0$$
$$(7.153)$$

Here, and in all following considerations of the second order homogeneous difference equation, we can and will (provided $d_2 \neq 0$) assume that $d_2 > 0$. We assume further that $d_1 < 0$. For the case that $d_1 > 0$, we can write $y(x) = e^{i\pi x}z(x)$. The difference equation for $z(x)$ is then

$$(c_2 + d_2(x + 2))z(x + 2) + (c_1' + d_1'(x + 1))z(x + 1) + (c_0 + d_0 x)z(x) = 0$$
$$(7.154)$$

so that $z(x)$ obeys a difference equation similar to that for $y(x)$, but in which the parameter $d_1' < 0$.

Referring to the second order homogeneous differential equation with linear coefficients, $(c_2 + d_2 x)y''(x) + (c_1 + d_1 x)y'(x) + (c_0 + d_0 x)y(x) = 0$, discussed in Sect. 7.4, it was seen that the solution can always be expressed in terms of the confluent hypergeometric function, taking on a particular form of that function, namely the Bessel function, for particular values of d_0, d_1 and d_2. A somewhat similar but not identical situation pertains to the second order homogeneous difference equation in that the functional form of the solution depends on the values of the parameters d_0, d_1 and d_2. However, for the *difference* equation the resulting solution is, depending on the values of these parameters, either a confluent hypergeometric function, $_1F_1$, or a hypergeometric function $_2F_1$. To classify the various choices of the parameters d_0, d_1 and d_2, we define the symbol $(\delta_0, \delta_1, \delta_2)$ in which $\delta_i = 0$ if $d_i = 0$ and $\delta_i = 1$ if $d_i \neq 0$, $(i = 0, 1, 2)$. We then consider the nine possibilities: $(1, 1, 1)$ with $d_1^2 - 4d_0 d_2 \neq 0$, $(1, 1, 1)$ with $d_1^2 - 4d_0 d_2 = 0, (1, 0, 1), (1, 1, 0), (0, 1, 1), (0, 1, 0), (0, 0, 1), (1, 0, 0), (0, 0, 0)$. In the last case, $(0, 0, 0)$, we have a difference equation with constant coefficients, given here since it includes the Tchebicheff polynomials, $T_n(x)$ and $U_n(x)$. We will derive explicit solutions to the equation defined by each of these sets of parameters. As we will show, however, some of these equations have solutions which are related simply by a multiplicative function. Thus, $(0, 1, 1)$ and $(1, 1, 0)$ are so related, as are $(0, 0, 1)$ and $(1, 0, 0)$. For the parameters given by $(1, 1, 0), (1, 0, 0)$ and $(0, 1, 0)$, as well as for $(1, 1, 1)$ with $d_1^2 - 4d_0 d_2 = \Delta^2 = 0$, the solution is expressed in terms of the confluent hypergeometric function $_1F_1$ or a particular form of that function—the Bessel function or the parabolic cylinder function. For the two remaining sets of parameters, $(1, 1, 1)$ with $\Delta \neq 0$ and $(1, 0, 1)$, the solutions are expressed in terms of the hypergeometric function $_2F_1$. These include the Legendre polynomials $P_n(x)$ and $Q_n(x)$, the Legendre functions $P_\nu^\mu(z)$ and $Q_\nu^\mu(z)$, the Gegenbauer polynomials $C_n^\nu(x)$, the Gegenbauer functions $C_\alpha^\nu(x)$ and $D_\alpha^\nu(x)$, the associated Legendre polynomials $P_n^{-\alpha}(z)$, the hypergeometric polynomials $_2F_1(-n, b; c; x)$ and the hypergeometric function $_2F_1(a, b; c; x)$ in which the recursion is in either a or b. These results may be expressed succinctly by the statement that if $d_0d_2\Delta \neq 0$ then the solution is expressed by the hypergeometric function $_2F_1$; if $d_0d_2\Delta = 0$ then the solution is expressed in terms of the confluent hypergeometric function $_1F_1$.

We first consider the cases (1, 1, 1) and (1, 0, 1) (i.e., $d_0 \neq 0$, $d_2 \neq 0$), and $d_1^2 - 4d_0d_2 \neq 0$, in which case the roots of $f_1(t) = 0$ are distinct. We can then write

$$f_1(t) = d_2(\alpha_1 - t)(\alpha_2 - t) \tag{7.155}$$

in which

$$\alpha_1 = \frac{-d_1 + \Delta}{2d_2}, \quad \alpha_2 = \frac{-d_1 - \Delta}{2d_2}, \quad \frac{1}{\alpha_1} = \frac{-d_1 - \Delta}{2d_0}, \quad \frac{1}{\alpha_2} = \frac{-d_1 + \Delta}{2d_0}$$

$$\Delta^2 = d_1^2 - 4d_0d_2, \quad \alpha_1 - \alpha_2 = \frac{\Delta}{d_2}, \quad \alpha_1\alpha_2 = \frac{d_0}{d_2} \tag{7.156}$$

Further, since we can consider $d_2 > 0$ and $d_1 \leq 0$, we note that

$$0 < \frac{\alpha_2}{\alpha_1} < 1 \quad \text{for } \Delta^2 > 0 \text{ and } d_0 > 0$$

$$-1 < \frac{\alpha_2}{\alpha_1} < 0 \quad \text{for } \Delta^2 > 0 \text{ and } d_0 < 0$$

$$\frac{\alpha_2}{\alpha_1} = e^{-2i\phi} \quad \text{for } 4d_0d_2 > d_1^2 \tag{7.157}$$

where

$$\tan \phi = \left| \frac{\Delta}{d_1} \right| \tag{7.158}$$

Then from (7.147), (7.149), (7.155) and (7.156) with $N = 2$,

$$\frac{v'(t)}{v(t)} = \frac{f_0(t)}{tf_1(t)} = \frac{f_0(t)}{d_2t(\alpha_1 - t)(\alpha_2 - t)} = \frac{\beta_0}{t} + \frac{\beta_1}{\alpha_1 - t} + \frac{\beta_2}{\alpha_2 - t}$$

$$= \frac{\beta_0}{t} + \frac{\beta_1}{\alpha_1 - t} - \frac{\beta_2}{t - \alpha_2} \tag{7.159}$$

where

$$f_0(t) = c_0 + c_1t + c_2t^2 \tag{7.160}$$

and

$$\beta_0 = \frac{f_0(0)}{d_2 \alpha_1 \alpha_2} = \frac{c_0}{d_0}$$

$$\beta_1 = \frac{f_0(\alpha_1)}{d_2 \alpha_1 (\alpha_2 - \alpha_1)} = -\frac{f_0(\alpha_1)}{\alpha_1 \Delta} \qquad (7.161)$$

$$\beta_2 = \frac{f_0(\alpha_2)}{d_2 \alpha_2 (\alpha_1 - \alpha_2)} = \frac{f_0(\alpha_2)}{\alpha_2 \Delta}$$

We note the relation of the exponents β_1 and β_2 to the exponents γ_1 and γ_2 derived for the corresponding second order differential equation, given in (7.97):

$$\beta_1 = -\frac{\gamma_1}{\alpha_1}, \quad \beta_2 = -\frac{\gamma_2}{\alpha_2} \qquad (7.162)$$

As is seen in Chap. 8, section "Difference Equations with Linear Coefficients", the classical functions which are solutions of second order homogeneous difference equations with linear coefficients are characterized by the particular combination of coefficients

$$\frac{c_0}{d_0} - 2\frac{c_1}{d_1} + \frac{c_2}{d_2} \qquad (7.163)$$

Indeed, difference equations for which $\frac{c_0}{d_0} - 2\frac{c_1}{d_1} + \frac{c_2}{d_2} = 0$ (and hence, from (7.164), $\beta_1 = \beta_2 = \frac{1}{2}(\frac{c_0}{d_0} - \frac{c_2}{d_2})$) have solutions which can be expressed in terms of the Legendre functions $P_\nu^\mu(z)$ and $Q_\nu^\mu(z)$; (this includes the Gegenbauer functions $C_\alpha^{(\nu)}(z)$ and $D_\alpha^{(\nu)}(z)$, which can be expressed in terms of the Legendre functions $P_\nu^\mu(z)$ and $Q_\nu^\mu(z)$, for which see Eqs. (8.45)–(8.48). We therefore express β_1 and β_2 as given above in (7.161) in terms of this expression. From (7.147), (7.155) and (7.156) we have[2]

$$\beta_1 = \frac{d_1}{2\Delta}\left[\frac{c_0}{d_0} - \frac{2c_1}{d_1} + \frac{c_2}{d_2}\right] + \frac{1}{2}\left(\frac{c_0}{d_0} - \frac{c_2}{d_2}\right)$$

$$\beta_2 = -\frac{d_1}{2\Delta}\left[\frac{c_0}{d_0} - \frac{2c_1}{d_1} + \frac{c_2}{d_2}\right] + \frac{1}{2}\left(\frac{c_0}{d_0} - \frac{c_2}{d_2}\right) \qquad (7.164)$$

$$\beta_0 = \frac{c_0}{d_0}$$

so that from (7.152) and (7.159) we have

$$v(t) = A t^{\beta_0} (\alpha_1 - t)^{-\beta_1} (\alpha_2 - t)^{-\beta_2} \qquad (7.165)$$

and, from (7.137),

$$y(x) = A \int t^{\beta_0 + x - 1} (\alpha_1 - t)^{-\beta_1} (\alpha_2 - t)^{-\beta_2} \, dt \qquad (7.166)$$

---

[2]For $d_1 = 0$ we have $\beta_1 = -\frac{c_1}{\Delta} + \frac{1}{2}\left(\frac{c_0}{d_0} - \frac{c_2}{d_2}\right)$, $\beta_2 = \frac{c_1}{\Delta} + \frac{1}{2}\left(\frac{c_0}{d_0} - \frac{c_2}{d_2}\right)$.

The integration path must now be chosen to satisfy the condition given in (7.150), namely

$$\left. f_1(t)t^x v(t)\right| = 0, \tag{7.167}$$

the particular choice depending on the values of $\beta_0 + x$, β_1 and β_2. This condition is satisfied if we assume for the moment that $\beta_0 + x > 0$, $\beta_1 < 1$, $\beta_2 < 1$, in which case each of the following three open integration paths gives a solution $y(x)$: Noting from (7.157) that $\left|\frac{\alpha_2}{\alpha_1}\right| \le 1$, we consider the following two solutions to the difference equation (7.153): (We show later that these two solutions are linearly independent.)

$$y_1(x) = A \int_{\alpha_2}^{\alpha_1} t^{\beta_0 + x - 1} (\alpha_1 - t)^{-\beta_1} (t - \alpha_2)^{-\beta_2} \, dt$$

$$y_2(x) = A \int_0^{\alpha_2} t^{\beta_0 + x - 1} (\alpha_1 - t)^{-\beta_1} (\alpha_2 - t)^{-\beta_2} \, dt \tag{7.168}$$

In each of these integrals, a simple transformation of integration variable will express the integral in the form commonly used for the hypergeometric function, viz.,

$$_2F_1(a, b; c; z) = \frac{\Gamma(c)}{\Gamma(b)\Gamma(c-b)} \int_0^1 t^{b-1}(1-t)^{c-b-1}(1-zt)^{-a} \, dt \tag{7.169}$$

In the integral for $y_1(x)$, the transformation of integration variable $t = \alpha_1 - (\alpha_1 - \alpha_2)s$ gives

$$y_1(x) = \alpha_1^{\beta_0 + x - 1}(\alpha_1 - \alpha_2)^{1 - \beta_1 - \beta_2} A \int_0^1 \frac{s^{-\beta_1}(1-s)^{-\beta_2}}{(1-zs)^{1-\beta_0-x}} \, ds, \qquad z = 1 - \frac{\alpha_2}{\alpha_1}$$

$$= \alpha_1^{\beta_0 + x - 1}(\alpha_1 - \alpha_2)^{1 - \beta_1 - \beta_2} A \frac{\Gamma(1-\beta_1)\Gamma(1-\beta_2)}{\Gamma(2-\beta_1-\beta_2)} {}_2F_1\left(1 - \beta_0 - x, 1 - \beta_1; 2 - \beta_1 - \beta_2; 1 - \frac{\alpha_2}{\alpha_1}\right) \tag{7.170}$$

In the integral for $y_2(x)$, the transformation of integration variable $t = \alpha_2 s$ gives

$$y_2(x) = \alpha_2^{\beta_0 + x - \beta_2} \alpha_1^{-\beta_1} A \int_0^1 \frac{s^{\beta_0 + x - 1}(1-s)^{-\beta_2}}{(1-zs)^{\beta_1}} \, ds, \qquad z = \frac{\alpha_2}{\alpha_1}$$

$$= \alpha_2^{\beta_0 + x - \beta_2} \alpha_1^{-\beta_1} A \frac{\Gamma(\beta_0 + x)\Gamma(1-\beta_2)}{\Gamma(\beta_0 + x + 1 - \beta_2)} {}_2F_1\left(\beta_1, \beta_0 + x; \beta_0 + x + 1 - \beta_2; \frac{\alpha_2}{\alpha_1}\right) \tag{7.171}$$

A judicious choice for A in $y_1(x)$ and $y_2(x)$ permits us to define these two solutions so that there are now no restrictions on β_1 and β_2, and that the only remaining restriction is $\beta_0 + x \ne 0, -1, -2, \ldots$ in $y_2(x)$. We can therefore define

$$y_1(x) = \alpha_1^{\beta_0 + x} \frac{1}{\Gamma(2 - \beta_1 - \beta_2)} {}_2F_1\left(1 - \beta_0 - x, 1 - \beta_1; 2 - \beta_1 - \beta_2; 1 - \frac{\alpha_2}{\alpha_1}\right) \tag{7.172}$$

and

$$y_2(x) = \alpha_2^{\beta_0+x} \frac{\Gamma(\beta_0+x)}{\Gamma(\beta_0+x+1-\beta_2)} \, {}_2F_1\left(\beta_1, \beta_0+x; \beta_0+x+1-\beta_2; \tfrac{\alpha_2}{\alpha_1}\right)$$
(7.173)

In the definition just given for $y_1(x)$ the argument of the hypergeometric function is <1 for $0 < \alpha_2/\alpha_1 < 1$. A further consideration of $y_1(x)$ depending on the values of the parameters β_0+x, β_1, β_2 and α_2/α_1 will be given shortly.

We now show that $y_1(x)$ and $y_2(x)$ are linearly independent solutions of the difference equation (7.153); that is, that there is no factor A, independent of x, such that $y_1(x) = Ay_2(x)$. To this end we examine the asymptotic approximations for $y_1(x)$ and $y_2(x)$ and find that their respective asymptotic approximations are manifestly not linearly related. The asymptotic approximations are most easily obtained by transforming the hypergeometric functions in $y_1(x)$ and $y_2(x)$ so that the large parameter x appears only in the third place in the hypergeometric function, in which case the hypergeometric function goes to unity as $x \to \infty$ (see [40, Sect. 5.5]). The asymptotic approximation for $y_2(x)$ is obtained straightforwardly. Making the transformation, (H.2),

$$\,{}_2F_1\left(\beta_1, \beta_0+x; \beta_0+x+1-\beta_2; \tfrac{\alpha_2}{\alpha_1}\right) = \left(1 - \tfrac{\alpha_2}{\alpha_1}\right)^{-\beta_1} {}_2F_1\left(\beta_1, 1-\beta_2; \beta_0+x+1-\beta_2; \tfrac{\alpha_2}{\alpha_2-\alpha_1}\right)$$
(7.174)

we note that for $x \gg 1$, (see [40, Sect. 5.5]),

$$\,{}_2F_1\left(\beta_1, 1-\beta_2; \beta_0+x+1-\beta_2; \tfrac{\alpha_2}{\alpha_2-\alpha_1}\right) = 1 + O\left(\tfrac{1}{x}\right)$$
(7.175)

from which

$$y_2(x) = \alpha_2^{\beta_0+x} \frac{\Gamma(\beta_0+x)}{\Gamma(\beta_0+x+1-\beta_2)} \left(1 - \tfrac{\alpha_2}{\alpha_1}\right)^{-\beta_1} \left(1 + O\left(\tfrac{1}{x}\right)\right)$$
(7.176)

For $y_1(x)$ we first make the transformation

$$\,{}_2F_1\left(1-\beta_0-x, 1-\beta_1; 2-\beta_1-\beta_2; 1-\tfrac{\alpha_2}{\alpha_1}\right)$$
$$= \left(\tfrac{\alpha_2}{\alpha_1}\right)^{\beta_0+x-\beta_2} {}_2F_1\left(\beta_0+x+1-\beta_1-\beta_2, 1-\beta_2; 2-\beta_1-\beta_2; 1-\tfrac{\alpha_2}{\alpha_1}\right)$$
$$\overset{!}{=} \left(\tfrac{\alpha_2}{\alpha_1}\right)^{\beta_0+x-\beta_2} {}_2F_1\left(1-\beta_2, \beta_0+x+1-\beta_1-\beta_2; 2-\beta_1-\beta_2; 1-\tfrac{\alpha_2}{\alpha_1}\right)$$
(7.177)

and then follow with the transformation (see [13, Sect. 2.9, (26), p. 106] and [40, Sect. 4, (26)])

$$
{}_2F_1\left(1 - \beta_2, \beta_0 + x + 1 - \beta_1 - \beta_2; 2 - \beta_1 - \beta_2; 1 - \tfrac{\alpha_2}{\alpha_1}\right)
$$
$$
= e^{i\pi(1-\beta_2)}\frac{\Gamma(2 - \beta_1 - \beta_2)\Gamma(\beta_0 + x)}{\Gamma(1 - \beta_1)\Gamma(\beta_0 + x + 1 - \beta_2)}\left(1 - \tfrac{\alpha_2}{\alpha_1}\right)^{\beta_2 - 1}
$$
$$
\times {}_2F_1\left(1 - \beta_2, \beta_1; \beta_0 + x + 1 - \beta_2; \tfrac{\alpha_2}{\alpha_2 - \alpha_1}\right)
$$
$$
+ \frac{\Gamma(2 - \beta_1 - \beta_2)\Gamma(\beta_0 + x)}{\Gamma(1 - \beta_2)\Gamma(\beta_0 + x + 1 - \beta_1)}\left(1 - \tfrac{\alpha_2}{\alpha_1}\right)^{\beta_1 - 1}\left(\tfrac{\alpha_2}{\alpha_1}\right)^{-\beta_0 - x + \beta_2}
$$
$$
\times {}_2F_1\left(\beta_2, 1 - \beta_1; \beta_0 + x + 1 - \beta_1; \tfrac{\alpha_1}{\alpha_1 - \alpha_2}\right) \tag{7.178}
$$

from which

$$
y_1(x) = \alpha_2^{\beta_0 + x} e^{i\pi(1-\beta_2)}\left(1 - \tfrac{\alpha_2}{\alpha_1}\right)^{\beta_2 - 1}\left(\tfrac{\alpha_1}{\alpha_2}\right)^{\beta_2}\frac{\Gamma(\beta_0 + x)}{\Gamma(1 - \beta_1)\Gamma(\beta_0 + x + 1 - \beta_2)}
$$
$$
\times {}_2F_1\left(1 - \beta_2, \beta_1; \beta_0 + x + 1 - \beta_2; \tfrac{\alpha_2}{\alpha_2 - \alpha_1}\right)
$$
$$
+ \alpha_1^{\beta_0 + x}\left(1 - \tfrac{\alpha_2}{\alpha_1}\right)^{\beta_1 - 1}\frac{\Gamma(\beta_0 + x)}{\Gamma(1 - \beta_2)\Gamma(\beta_0 + x + 1 - \beta_1)}
$$
$$
\times {}_2F_1\left(\beta_2, 1 - \beta_1; \beta_0 + x + 1 - \beta_1; \tfrac{\alpha_1}{\alpha_1 - \alpha_2}\right) \tag{7.179}
$$

From (7.173) and (7.174) we note that the first term in (7.179) is a multiple of $y_2(x)$. For $x \gg 1$ we have

$$
{}_2F_1\left(1 - \beta_2, \beta_1; \beta_0 + x + 1 - \beta_2; \tfrac{\alpha_2}{\alpha_2 - \alpha_1}\right) = 1 + O\left(\tfrac{1}{x}\right)
$$
$$
{}_2F_1\left(\beta_2, 1 - \beta_1; \beta_0 + x + 1 - \beta_1; \tfrac{\alpha_1}{\alpha_1 - \alpha_2}\right) = 1 + O\left(\tfrac{1}{x}\right) \tag{7.180}
$$

from which, for $x \gg 1$, neglecting terms of relative order $1/x$,

$$
y_1(x) = \alpha_2^{\beta_0 + x} e^{i\pi(1-\beta_2)}\left(1 - \tfrac{\alpha_2}{\alpha_1}\right)^{\beta_2 - 1}\left(\tfrac{\alpha_1}{\alpha_2}\right)^{\beta_2}\frac{\Gamma(\beta_0 + x)}{\Gamma(1 - \beta_1)\Gamma(\beta_0 + x + 1 - \beta_2)}
$$
$$
+ \alpha_1^{\beta_0 + x}\left(1 - \tfrac{\alpha_2}{\alpha_1}\right)^{\beta_1 - 1}\frac{\Gamma(\beta_0 + x)}{\Gamma(1 - \beta_2)\Gamma(\beta_0 + x + 1 - \beta_1)} \tag{7.181}
$$

From (7.157) we note that provided $d_1^2 > 4d_0d_2$, we have $|\alpha_2| < |\alpha_1|$, so that for $x \gg 1$ we may neglect the term in $y_1(x)$ with the factor $\alpha_2^{\beta_0 + x}$, which is equal to $y_2(x)$ apart from a factor independent of x. We then have, expanding the gamma functions for $x \gg 1$ and neglecting terms of relative order $1/x$,

$$y_2(x) = \alpha_2^{\beta_0+x} \left(1 - \frac{\alpha_2}{\alpha_1}\right)^{-\beta_1} (x + \beta_0)^{\beta_2-1}$$

$$y_1(x) = \alpha_1^{\beta_0+x} \frac{\left(1 - \frac{\alpha_2}{\alpha_1}\right)^{\beta_1-1}}{\Gamma(1 - \beta_2)} (x + \beta_0)^{\beta_1-1} \tag{7.182}$$

The asymptotic approximation given here[3] for $y_1(x)$ is identical to that given in [39, Eq. (2.6), p. 372]. For $4d_1^2 > d_0 d_2$, for which $|\alpha_2| < |\alpha_1|$, the solutions $y_1(x)$ and $y_2(x)$ are thus clearly linearly independent. For $4d_1^2 < d_0 d_2$, $\alpha_2 = \alpha_1^*$ from (7.156), so that $|\alpha_2| = |\alpha_1|$. The asymptotic approximation for $y_1(x)$ in this case is given in [39, Eqs. (2.14) and (2.9), pp. 372, 373], where a uniform asymptotic expansion of the hypergeometric function is derived. However, the dominant behavior for $x \gg 1$ is still in the factor $\alpha_1^{\beta_0+x}$, so that $y_1(x)$ and $y_2(x)$ are again linearly independent.

We now return to a consideration of $y_1(x)$ as given in (7.172). If either $\beta_0 + x = 1, 2, \ldots$ or $\beta_1 = 1, 2, \ldots$ then $y_1(x)$ is a polynomial, and the hypergeometric function in this expression is finite. With the transformation given in (7.177) it follows that $y_1(x)$ is also a polynomial if $\beta_2 = 1, 2, \ldots$ If none of these three conditions is satisfied and if in addition $\alpha_2/\alpha_1 < 0$, then the transformation given in (7.179) expresses $y_1(x)$ in terms of hypergeometric functions with an argument z such that $0 < z < 1$. However, as noted, the term in (7.179) with factor $\alpha_2^{\beta_0+x}$ is a multiple of $y_2(x)$. The second term in (7.179), with factor $\alpha_1^{\beta_0+x}$, is therefore also a solution of the difference equation. Therefore, for $\alpha_2/\alpha_1 < 0$ we can define the solution $y_1(x)$ by this second term. Throughout, in making the following definitions we are free to omit any factors which do not depend on x. In summary, then, we define

$$y_1(x) = \alpha_1^{\beta_0+x} \frac{1}{\Gamma(2 - \beta_1 - \beta_2)} \, {}_2F_1 \left(1 - \beta_0 - x, 1 - \beta_1; 2 - \beta_1 - \beta_2; 1 - \frac{\alpha_2}{\alpha_1}\right) \tag{7.183}$$

if either $\beta_0 + x = 1, 2, \ldots$, or $\beta_1 = 1, 2, \ldots$ (in which case $y_1(x)$ is a polynomial), or if $0 < \alpha_2/\alpha_1 < 1$ (i.e., $0 < 4d_0 d_2 < d_1^2$).

From (7.177) we define

$$y_1(x) = \alpha_2^{\beta_0+x} \frac{1}{\Gamma(2 - \beta_1 - \beta_2)} \, {}_2F_1 \left(1 - \beta_2, \beta_0 + x + 1 - \beta_1 - \beta_2; 2 - \beta_1 - \beta_2; 1 - \frac{\alpha_2}{\alpha_1}\right) \tag{7.184}$$

if $\beta_2 = 1, 2, \ldots$ (in which case $y_1(x)$ is again a polynomial).

Finally, if $-1 < \alpha_2/\alpha_1 < 0$, i.e., $d_0 < 0$, (see (7.157)), from (7.179), omitting factors which are independent of x, we define

$$y_1(x) = \alpha_1^{\beta_0+x} \frac{\Gamma(\beta_0 + x)}{\Gamma(\beta_0 + x + 1 - \beta_1)} \, {}_2F_1 \left(\beta_2, 1 - \beta_1; \beta_0 + x + 1 - \beta_1; \frac{\alpha_1}{\alpha_1 - \alpha_2}\right) \tag{7.185}$$

---

[3]As noted in Appendix A of [22], Eqs. (13)–(15) in [13] are incorrect.

This last expression for $y_1(x)$ is appropriate for the consideration of the other set of parameters for which the solution to the difference equation is a hypergeometric function, namely when $d_1 = 0$, denoted earlier by $(1, 0, 1)$. In that case, from (7.156) we have $\alpha_2 = -\alpha_1$, and from (7.164),

$$\beta_1 = -\frac{c_1}{\Delta} + \frac{1}{2}\left(\frac{c_0}{d_0} - \frac{c_2}{d_2}\right), \qquad \beta_2 = \frac{c_1}{\Delta} + \frac{1}{2}\left(\frac{c_0}{d_0} - \frac{c_2}{d_2}\right) \qquad (7.186)$$

$$y_1(x) = \alpha_1^{\beta_0+x} \frac{\Gamma(\beta_0 + x)}{\Gamma(\beta_0 + x + 1 - \beta_1)} \, {}_2F_1\left(\beta_2, 1 - \beta_1; \beta_0 + x + 1 - \beta_1; \tfrac{1}{2}\right) \quad (7.187)$$

For $y_2(x)$, from (7.173) and making the transformation given in (7.174) with $\alpha_2 = -\alpha_1$ and β_1, and β_2 given in (7.186), we have, again neglecting factors independent of x,

$$y_2(x) = e^{i\pi(\beta_0+x)} \alpha_1^{\beta_0+x} \frac{\Gamma(\beta_0 + x)}{\Gamma(\beta_0 + x + 1 - \beta_2)} \, {}_2F_1\left(\beta_1, 1 - \beta_2; \beta_0 + x + 1 - \beta_2; \tfrac{1}{2}\right)$$
$$(7.188)$$

We now consider those sets of parameters for which the solution to the difference equation is expressed in terms of the confluent hypergeometric function or one of its particular forms, namely, the Bessel function or the parabolic cylinder function. We first consider the case denoted earlier by $(1, 1, 1)$ (i.e., $d_0 d_1 d_2 \neq 0$) but with $d_1^2 - 4d_0 d_2 = 0$. We now have, from (7.155) and (7.156),

$$\alpha_1 = \alpha_2 \equiv \alpha = -\frac{d_1}{2d_2} > 0, \quad \frac{1}{\alpha} = -\frac{d_1}{2d_0}, \quad d_2\alpha^2 = d_0 \qquad (7.189)$$

and

$$f_1(t) = d_2(\alpha - t)^2 \qquad (7.190)$$

In place of (7.159) we can write

$$\frac{v'(t)}{v(t)} = \frac{f_0(t)}{tf_1(t)} = \frac{f_0(t)}{td_2(\alpha - t)^2} = \frac{\beta_0}{t} + \frac{\beta_1}{\alpha - t} + \frac{\beta_2}{(\alpha - t)^2} \qquad (7.191)$$

in which, from (7.160), $f_0(t) = c_0 + c_1 t + c_2 t^2$. Equating leading terms in (7.191), we then have: As $t \to 0$,

$$\beta_0 = \frac{c_0}{d_2\alpha^2} = \frac{c_0}{d_0}; \qquad (7.192)$$

as $t \to \infty$,

$$\beta_0 - \beta_1 = \frac{c_2}{d_2}, \quad \text{so that} \quad \beta_1 = \frac{c_0}{d_0} - \frac{c_2}{d_2}; \qquad (7.193)$$

and as $t \to \alpha$,

$$\beta_2 = \frac{f_0(\alpha)}{\alpha d_2} = \alpha \left(\frac{c_0}{d_0} - 2\frac{c_1}{d_1} + \frac{c_2}{d_2} \right) \tag{7.194}$$

Integrating (7.191) gives

$$v(t) = At^{\beta_0}(\alpha - t)^{-\beta_1} e^{\frac{\beta_2}{\alpha - t}} \tag{7.195}$$

from which

$$y(x) = \int t^{x-1} v(t)\, dt = A \int t^{\beta_0 + x - 1}(\alpha - t)^{-\beta_1} e^{\frac{\beta_2}{\alpha - t}}\, dt \tag{7.196}$$

in which we must choose an integration path such that

$$f_1(t)t^x v(t)\Big| = 0 \tag{7.197}$$

In order to express the solution $y(x)$ as a commonly used integral representation for the confluent hypergeometric function, we consider different transformations of the integration variable in (7.196), one leading to functions of the form $_1F_1(a; c; z)$, the others to functions of the form $U(a; c; z)$. Then, depending on the particular choice of integration path, different expressions are obtained for the confluent hypergeometric function. Which of these expressions provides a useful expression for the solution $y(x)$ depends, as will be seen, on the values of the parameters $\beta_0, \beta_1, \beta_2$ and the variable z. To that end we first simplify the integral for $y(x)$, (7.196), substituting αt for t in the integrand and neglecting factors independent of x, giving

$$y(x) = \alpha^{\beta_0 + x} \int t^{\beta_0 + x - 1}(1 - t)^{-\beta_1} e^{\frac{z}{1-t}}\, dt \tag{7.198}$$

where

$$z = \frac{\beta_2}{\alpha} \tag{7.199}$$

The transformation of integration variable $t = \frac{s-1}{s}$ in (7.198) leads to functions of the form $_1F_1(a; c; z)$, giving[4]

$$y(x) = \alpha^{\beta_0 + x} e^{\pm i\pi(\beta_0 + x - 1)} \int s^{\beta_1 - \beta_0 - x - 1}(1 - s)^{\beta_0 + x - 1} e^{zs}\, ds \tag{7.200}$$

---

[4]The alternate transformation, $t = \frac{s}{s-1}$, leads to functions which follow from those obtained from $t = \frac{s-1}{s}$ using Kummer's transformation [36, Sect. 13.2(vii), Eq. 13.2.39].

Alternatively, it will be useful to write this integral in the form

$$y(x) = \alpha^{\beta_0 + x} \int s^{-(2-\beta_1)} (1 - \tfrac{1}{s})^{\beta_0 + x - 1} e^{zs} \, ds \qquad (7.201)$$

The particular form of the function $_1F_1(a; c; z)$ defined by these integrals depends on the choice of integration path, and will be considered shortly (see, e.g., [36, Sects. 13.4 (i) and (ii)]).

The transformations of integration variable leading to confluent hypergeometric functions of the form $U(a; c; z)$ are $t = \frac{s+1}{s}$ and $t = \frac{s}{s+1}$. Writing $t = \frac{s+1}{s}$ we have

$$y(x) = \alpha^{\beta_0 + x} e^{i\pi(\beta_1 - 1)} \int s^{\beta_1 - \beta_0 - x - 1} (1 + s)^{\beta_0 + x - 1} e^{-zs} \, ds \qquad (7.202)$$

Writing $t = \frac{s}{s+1}$ we have

$$y(x) = \alpha^{\beta_0 + x} e^{z} \int s^{\beta_0 + x - 1} (1 + s)^{\beta_1 - \beta_0 - x - 1} e^{zs} \, ds \qquad (7.203)$$

Again, the particular form of the function $U(a; c; z)$ defined by these integrals depends on the choice of integration path (see [36, Sects. 13.4 (i) and (ii)]).

In these integrals, the relation of the parameters $\beta_0 + x - 1 = \frac{c_0}{d_0} + x - 1$ and $\beta_0 - \beta_1 + x + 1 = \frac{c_2}{d_2} + x + 1$ to the coefficients of the difference equation (7.153) is seen most clearly if we define

$$y(x + 1) = \alpha^x w(x) \qquad (7.204)$$

The difference equation for $w(x)$ is then,[5] from (7.153) and (7.189),

$$(\gamma_2 + x + 1) w(x + 1) - 2(\gamma_1 + x) w(x) + (\gamma_0 + x - 1) w(x - 1) = 0 \qquad (7.205)$$

in which we define

$$\gamma_i = \frac{c_i}{d_i}, \qquad i = 0, 1, 2 \qquad (7.206)$$

It is important to note that if we assume to have two independent initial conditions, for example $w(0)$ and $w(1)$, then the recursion defined by (7.205) fails if $\gamma_2 + x + 1 = 0$ in that $w(x + 1)$ is then not determined.

From (7.198)–(7.203) we derive, in Appendix I, solutions to (7.205) corresponding to particular choices of integration paths. The solutions $w(x)$, divested of factors independent of x, are given in Eqs. (I.12)–(I.29).

---

[5] We note that, in terms of the difference operator Δ defined in Chap. 1, the difference equation for $w(x)$ can be written in the form $(\gamma_2 + x + 1)\Delta^2 w(x - 1) + (2 - \beta_1)\Delta w(x - 1) + zw(x) = 0$, corresponding to the confluent hypergeometric function F5t given in Appendix I.

In order to obtain a solution to the difference equation that satisfies arbitrary initial conditions $y(x_0)$ and $y(x_0 + 1)$, we require two linearly independent solutions of the difference equation, $y_1(x)$ and $y_2(x)$. We therefore wish to chose, from among the solutions given in Appendix I, pairs of solutions which are linearly independent. The particular pair of solutions chosen will depend on the value of the parameters $\gamma_2 + x + 1$ and $\gamma_0 + x - 1$ in (7.205) as well as on $\beta_1 = \gamma_0 - \gamma_2$ and $z = \gamma_0 - 2\gamma_1 + \gamma_2$. The condition that the two solutions be linearly independent is that their Casoratian, $\mathcal{C}(x)$, be non-zero: $\mathcal{C}(x) = y_1(x)\, y_2(x + 1) - y_1(x + 1)\, y_2(x) \neq 0$. Noting that the Wronskians of the various solutions that we are considering are relatively well-known, (see, e.g., [36, Sect. 13.2(vi)]), we determine the Casoratian by expressing it in terms of the Wronskian by the use of raising and lowering operators; these relate the differential properties of the variable z in the solution to the discrete properties of the parameter x.

We note that all of the confluent hypergeometric functions listed in Appendix I are of the form $F(a + x; c; z)$ or $F(a - x; c; z)$, where F is either $_1F_1$ or U. We then make use of the raising operators

$$\frac{d}{dz}\left(z^{a+x}\, _1F_1(a + x; c; z)\right) = (a + x)z^{a+x-1}\, _1F_1(a + x + 1; c; z)$$

$$\frac{d}{dz}\left(z^{a+x} U(a + x; c; z)\right) = (a + x)(a + x - c + 1)z^{a+x-1} U(a + x + 1; c; z)$$

$$(7.207)$$

and the lowering operators

$$\frac{d}{dz}\left(e^{-z}z^{c-a+x}\, _1F_1(a - x; c; z)\right) = (c - a + x)e^{-z}z^{c-a+x-1}\, _1F_1(a - x - 1; c; z)$$

$$\frac{d}{dz}\left(e^{-z}z^{c-a+x} U(a - x; c; z)\right) = -e^{-z}z^{c-a+x-1} U(a - x - 1; c; z) \qquad (7.208)$$

The linearly independent pairs of solutions of the difference equation (7.205) and their corresponding region of validity are given below in Table 7.1, followed by a detailed derivation of the Casoratian for one of the pairs; the derivation for the remaining pairs is quite similar. The functions $U1$–$U4t$ and $F1$–$F5t$ are defined in Appendix I, Eqs. (I.22)–(I.29) and (I.12)–(I.21). It is to be observed that within each pair of linearly independent solutions, the variable and parameters are the same in the functions U and $_1F_1$. We note also that if one of the parameters $\gamma_2 + x + 2$ and $\gamma_0 + x$ is an integer while the other is not, then the second parameter in the confluent hypergeometric function, β_1 or $2 - \beta_1$, is not an integer. The gamma functions which multiply the functions U and $_1F_1$ enable one to use the raising and lowering operators to express the Casoratian in terms of the Wronskian. We assume throughout that $\gamma_2 + x + 1 \neq 0, -1, -2\ldots$ for any value of x in the range under consideration.

Table 7.1 Pairs of linearly independent solutions of the difference equation for $w(x)$

$\gamma_2 + x + 1 \neq 1, 2, 3, \ldots$		
$\gamma_0 + x = 1, 2, 3, \ldots$		
$z > 0$	(U2, F3)	
$z < 0$	(U3, F3t), (U3t, F5t)	
$\gamma_2 + x + 1 \neq 1, 2, 3, \ldots$		
$\gamma_0 + x = 0, -1, -2, \ldots$		
$z > 0$	(U2, F2), (U2t, F5)	
$z < 0$	(U4t, F5t)	
$\gamma_2 + x + 1 \neq 1, 2, 3, \ldots$		
$\gamma_0 + x \neq 0, \pm 1, \pm 2, \ldots$		
$z > 0,$	$\beta_1 \neq 0, -1, -2, \ldots,$	(U2, F3), (U2, F2)
$z > 0,$	$\beta_1 \neq 2, 3, 4, \ldots,$	(U2t, F5)
$z < 0,$	$\beta_1 \neq 0, -1, -2, \ldots,$	(U3, F3t)
$z < 0,$	$\beta_1 \neq 2, 3, 4, \ldots,$	(U4t, F5t), (U3t, F5t)
$\gamma_2 + x + 1 = 1, 2, 3, \ldots$		
$\gamma_0 + x \neq 1, 2, 3, \ldots$		
$z > 0$	(U2t, F5)	
$z < 0$	(U4t, F5t)	
$\gamma_2 + x + 1 = 1, 2, 3, \ldots$		
$\gamma_0 + x = 1, 2, 3, \ldots$		
$z > 0,$	see Appendix J	
$z < 0,$	$\beta_1 \neq 0, -1, -2, \ldots,$	(U3, F3t)
$z < 0,$	$\beta_1 \neq 2, 3, 4, \ldots,$	(U3t, F5t)

We give here the derivation of the Casoratian of the functions $U3$ and $F3t$, valid assuming $\gamma_0 + x \neq 0, -1, -2, \ldots$ and $\gamma_2 + x + 1 \neq 0, -1, -2, \ldots$ and $z < 0$. From (I.26) and (I.17), for conciseness of expression, we write the functions $U3$ and $F3t$ in the form

$$U3 = \Gamma(\gamma_0 + x)U(x, -z)$$

$$F3t = \frac{\Gamma(\gamma_0 + x)}{\Gamma(\gamma_2 + x + 1)}F(x, -z) \tag{7.209}$$

The Casoratian of $U3$ and $F3t$ is then

$$\mathcal{C}(U3, F3t) = \Gamma(\gamma_0 + x)U(x, -z)\frac{\Gamma(\gamma_0 + x + 1)}{\Gamma(\gamma_2 + x + 2)}F(x + 1, -z)$$

$$- \Gamma(\gamma_0 + x + 1)U(x + 1, -z)\frac{\Gamma(\gamma_0 + x)}{\Gamma(\gamma_2 + x + 1)}F(x, -z)$$

$$= \frac{\Gamma(\gamma_0 + x)\Gamma(\gamma_0 + x + 1)}{\Gamma(\gamma_2 + x + 1)}$$

$$\times \left[\frac{U(x, -z)}{(\gamma_2 + x + 1)} F(x + 1, -z) - U(x + 1, -z) F(x, -z) \right]$$

$$(7.210)$$

Here, using the raising operators,

$$U(x + 1) = \frac{z \frac{d}{dz} \left(z^{\gamma_0 + x} U(x, -z) \right)}{(\gamma_2 + x + 1)(\gamma_0 + x) z^{\gamma_0 + x}}$$

$$F(x + 1) = \frac{z \frac{d}{dz} \left(z^{\gamma_0 + x} F(x, -z) \right)}{(\gamma_0 + x) z^{\gamma_0 + x}}$$

$$(7.211)$$

from which

$$\mathcal{C}(U3, F3t) = \frac{\Gamma^2(\gamma_0 + x)}{\Gamma(\gamma_2 + x + 2)} z$$

$$\times \left[U(x, -z) \frac{d}{dz} F(x, -z) - F(x, -z) \frac{d}{dz} U(x, -z) \right]$$

$$= \frac{\Gamma^2(\gamma_0 + x)}{\Gamma(\gamma_2 + x + 2)} z$$

$$\times \mathcal{W}(U(x, -z), F(x, -z))$$

$$= -\frac{\Gamma(\gamma_0 + x)}{\Gamma(\gamma_2 + x + 2)} (-z)^{1 - \beta_1} e^{-z}$$

$$(7.212)$$

where $\mathcal{W}(U(x, -z), F(x, -z))$ is the Wronskian, given in [36, Sect. 13.2(vi), Eq. 13.2.34] and, from (7.194), (7.199) and (7.206),

$$z = \gamma_0 - 2\gamma_1 + \gamma_2 \qquad (7.213)$$

The case in which $z = 0$, which is not included in Table 7.1, is considered here. The difference equation (7.205) then has, neglecting factors independent of x throughout, one solution $w(x) = 1$ given by F5 or F5t, and a second linearly independent solution $w(x) = \Gamma(\gamma_0 + x)/\Gamma(\gamma_2 + x + 1)$, given by F3 or F3t if $\gamma_0 + x \neq 0, -1, -2\ldots$ or by F4 or F4t if $\gamma_0 + x \neq 1, 2, 3\ldots$ For integer values of x the second solution can be written in the form $w(n) = (\gamma_0)_n/(\gamma_2 + 1)_n$. For the case in which $\gamma_0 = \gamma_2 + 1$, the second solution may be written as a linear combination of the two solutions:

$$w(n) = \frac{(\gamma_2 + 1)}{(\gamma_0 - \gamma_2 - 1)} \left[\frac{(\gamma_0)_n}{(\gamma_2 + 1)_n} - \frac{\gamma_0}{\gamma_2 + 1} \right], \qquad (7.214)$$

from which $w(0) = -1$, $w(1) = 0$, and

$$w(n) = \gamma_0 \sum_{k=1}^{n-1} \frac{1}{\gamma_0 + k}, \qquad n \geq 2 \qquad (7.215)$$

7.6.1 Solution of Second Order Homogeneous Differential Equations with Linear Coefficients Through Transformation of Dependent and Independent Variables

Finally, we present an approach in which the dependent and independent variables of the differential equation are transformed so that the resulting equation is recognized to have a form whose solution is well-known. We carry through this approach for the second order homogeneous differential equation with coefficients linear in the independent variable. Explicit solutions are given in [13, Sect. 6.2, pp. 249–252], the transformed equation having solutions which are either a confluent hypergeometric function or a Bessel function (which may be expressed in terms of confluent hypergeometric functions), the specific form of the solution depending on the coefficients. A summary of transformations leading to the solution is given in [27, C. 2., Eq. (2.145), p. 434]. Here we present the derivation of the explicit solutions with a slight change of notation to facilitate comparison with the analysis given in the next section, where we apply this approach to the corresponding difference equation.

The specific function which is the solution to the second-order differential equation

$$(c_2 + d_2 x)y''(x) + (c_1 + d_1 x)y'(x) + (c_0 + d_0 x)y(x) = 0 \qquad (7.216)$$

depends on the value of the parameters d_0, d_1, and d_2, for which we use the classification discussed previously in Sect. 7.6 of this chapter in connection with the difference equation

$$(c_2 + d_2(x + 1))y(x + 1) + (c_1 + d_1 x)y(x) + (c_0 + d_0(x - 1))y(x - 1) = 0. \qquad (7.217)$$

There we defined the symbol $(\delta_0, \delta_1, \delta_2)$ in which $\delta_i = 0$ if $d_i = 0$ and $\delta_i = 1$ if $d_i \neq 0$, $(i = 0, 1, 2)$. We again consider the nine possibilities: $(1, 1, 1)$ with $d_1^2 - 4d_0 d_2 \neq 0$, $(1, 1, 1)$ with $d_1^2 - 4d_0 d_2 = 0$, $(1, 0, 1)$, $(1, 1, 0)$, $(0, 1, 1)$, $(0, 1, 0)$, $(0, 0, 1)$, $(1, 0, 0)$, $(0, 0, 0)$.

We start by making the change of dependent variable:

$$y(x) = e^{hx} w(x), \qquad (7.218)$$

the differential equation for $w(x)$ then being

$$(c_2 + d_2 x)w''(x) + [2(c_2 + d_2 x)h + (c_1 + d_1 x)]w'(x)$$
$$+ [(c_2 + d_2 x)h^2 + (c_1 + d_1 x)h + (c_0 + d_0 x)]w(x) = 0. \qquad (7.219)$$

The significant parameters in this equation are evident in the last term of this equation, specifically,

$$d_2h^2 + d_1h + d_0 \equiv \lambda(h), \qquad \Delta \equiv (d_1^2 - 4d_0d_2)^{\frac{1}{2}}$$
$$c_2h^2 + c_1h + c_0 \equiv \kappa(h), \qquad \gamma \equiv (c_1^2 - 4c_0c_2)^{\frac{1}{2}} \qquad (7.220)$$

We consider first the case $(0, 0, 0)$ $(d_2 = d_1 = d_0 = 0)$, for which the differential equation (7.219) then has constant coefficients:

$$c_2w''(x) + \kappa'(h)w'(x) + \kappa(h)w(x) = 0. \qquad (7.221)$$

Choosing h to be a solution of $\kappa(h) = 0$, $(h = h_\pm = \frac{-c_1 \pm \gamma}{2c_2})$, then gives the solution $w(x) =$ constant, from which we have the two linearly independent solutions to (7.216):

$$y_1(x) = e^{h_+ x}$$
$$y_2(x) = e^{h_- x} \qquad (7.222)$$

if $\gamma \neq 0$. If $\gamma = 0$ then $h_+ = h_- = -\frac{c_1}{2c_2}$ and $w''(x) = 0$, the second solution being $w(x) = x$, from which

$$y_2(x) = xe^{h_+ x} \qquad (7.223)$$

Next, considering $(1, 0, 0)$, $(d_0 \neq 0)$, the differential equation (7.219) is then

$$c_2w''(x) + \kappa'(h)w'(x) + (\kappa(h) + d_0x)w(x) = 0. \qquad (7.224)$$

Choosing $h = -\frac{c_1}{2c_2}$ so that $\kappa'(h) = 0$, we then have $\kappa(h) = -\frac{\gamma^2}{4c_2}$. The differential equation for $w(x)$ is then

$$w''(x) = \left(\frac{\gamma^2}{4c_2^2} - \frac{d_0}{c_2}x \right) w(x), \qquad (7.225)$$

which is essentially the differential equation for the Airy function (see [36, Chap. 9]). If we now make the change of independent variable

$$z = -(c_2d_0^2)^{-\frac{1}{3}}(\kappa(h) + d_0x), \qquad (7.226)$$

writing $w(x) = f(z)$, we have
$$f''(z) = zf(z) \qquad (7.227)$$

for which the two linearly independent solutions are $f(z) = Ai(z)$ and $f(z) = Bi(z)$, from which the solutions to equation (7.216) are

$$y_1(x) = e^{-\frac{c_1x}{2c_2}} Ai(z)$$
$$y_2(x) = e^{-\frac{c_1x}{2c_2}} Bi(z) \qquad (7.228)$$

If the argument of these Airy functions is positive, they can be expressed in terms of the modified Bessel functions $K_{\frac{1}{3}}$ and $I_{\frac{1}{3}}$. If the argument is negative they can be expressed in terms of the Bessel functions $J_{\frac{1}{3}}$ and $J_{-\frac{1}{3}}$, all of which can be expressed in terms of confluent hypergeometric functions, [36, Chap. 9].

Next we consider the remaining seven cases, for which either $d_2 \neq 0$ or $d_1 \neq 0$. We can then choose h to be a solution of the equation $\lambda(h) = 0$. The differential equation for $w(x)$, Eq. (7.219), is then

$$(c_2 + d_2 x)w''(x) + (\kappa'(h) + \lambda'(h)x)w'(x) + \kappa(h)w(x) = 0 \qquad (7.229)$$

If, in addition, $d_2 \neq 0$ then $\lambda'(h) = \Delta$ and Eq. (7.229) is

$$(c_2 + d_2 x)w''(x) + (\kappa'(h) + \Delta x)w'(x) + \kappa(h)w(x) = 0, \qquad (7.230)$$

which is essentially the differential equation for the confluent hypergeometric equation $_1F_1(a; c; x)$ provided $\Delta \neq 0$, since the factor of $w(x)$ is independent of x, [36, Chap. 13]. The two conditions, $d_2 \neq 0$ and $\Delta \neq 0$, are fulfilled for $(1, 1, 1)$ with $\Delta \neq 0$, $(1, 0, 1)$ and $(0, 1, 1)$, and for these three cases a linear transformation of the independent variable allows one to express the solution in terms of the confluent hypergeometric function: Defining

$$z = -\frac{\Delta}{d_2^2}(c_2 + d_2 x), \qquad (7.231)$$

writing $w(x) = f(z)$, we have

$$zf''(z) + (c - z)f'(z) - af(z) = 0 \qquad (7.232)$$

so that

$$f(z) = {}_1F_1(a; c; z) \qquad (7.233)$$

and hence

$$y(x) = e^{hx}\,{}_1F_1(a; c; z) \qquad (7.234)$$

in which

$$h = \frac{-d_1 + \Delta}{2d_2}$$

$$a = \frac{\kappa(h)}{\Delta}$$

$$c = \frac{c_1 d_2 - c_2 d_1}{d_2^2}$$

$$z = -\frac{\Delta}{d_2^2}(c_2 + d_2 x) \qquad (7.235)$$

We next consider $(1, 1, 0)$ and $(0, 1, 0)$, for which $d_2 = 0$, $d_1 \neq 0$. Then from $\lambda(h) = 0$ we have $h = -\frac{d_0}{d_1}$ and the differential equation (7.219) for $w(x)$ is

$$c_2 w''(x) + (\kappa'(h) + d_1 x)w'(x) + \kappa(h)w(x) = 0. \qquad (7.236)$$

If we now make the change of independent variable

$$z = -\frac{1}{2c_2 d_1}\left(\kappa'(h) + d_1 x\right)^2, \qquad (7.237)$$

writing $w(x) = f(z)$, it is seen that $f(z)$ satisfies the differential equation for the confluent hypergeometric function:

$$zf''(z) + (\tfrac{1}{2} - z)f'(z) - \frac{\kappa(h)}{2d_1}f(z) = 0. \qquad (7.238)$$

Thus, $f(z) = {}_1F_1(a; \tfrac{1}{2}; z)$, from which

$$y(x) = e^{hx}\,{}_1F_1(a; \tfrac{1}{2}; z) \qquad (7.239)$$

in which

$$h = -\frac{d_0}{d_1}$$
$$a = \frac{\kappa(h)}{2d_1}$$
$$z = -\frac{1}{2c_2 d_1}(\kappa'(h) + d_1 x)^2 \qquad (7.240)$$

Finally, we consider $(1, 1, 1)$ with $\Delta = 0$ and $(0, 0, 1)$, for which $d_2 \neq 0$ and $\Delta = 0$. Then, from $\lambda(h) = 0$, $h = -\frac{d_1}{2d_2}$. Equation (7.219) for $w(x)$ is then

$$(c_2 + d_2 x)w''(x) + \kappa'(h)w'(x) + \kappa(h)w(x) = 0. \qquad (7.241)$$

With the change of independent variable

$$z = \frac{2}{d_2}\kappa^{\frac{1}{2}}(c_2 + d_2 x)^{\frac{1}{2}}, \qquad (7.242)$$

writing $w(x) = f(z)$, the differential equation for $f(z)$ is

$$zf''(z) + (2\mu - 1)f'(z) + zf(z) = 0 \qquad (7.243)$$

in which $\mu = \frac{c_1 d_2 - c_2 d_1}{d_2^2}$. Then, making the transformation of dependent variable $g(z) = z^{1-\mu} f(z)$, we find that $g(z)$ obeys the differential equation for the Bessel function:

$$z^2 g''(z) + z g'(z) + (z^2 - (1 - \mu)^2) g(z) = 0, \qquad (7.244)$$

the solutions being $g(z) = J_{1-\mu}(z)$ and $g(z) = Y_{1-\mu}(z)$. Finally then, for the two cases $(1, 1, 1)$ with $\Delta = 0$ and $(0, 0, 1)$,

$$y(x) = e^{hx} z^{1-\mu} J_{1-\mu}(z) \qquad (7.245)$$

in which

$$h = -\frac{d_1}{2 d_2}$$

$$\mu = \frac{c_1 d_2 - c_2 d_1}{d_2^2}$$

$$z = \frac{2}{d_2} \kappa^{\frac{1}{2}}(h)(c_2 + d_2 x)^{\frac{1}{2}} \qquad (7.246)$$

We note that once one has made the change of dependent variable given in (7.218), defining the equation (7.219), all of the changes of independent variable introduced to produce differential equations in a well-recognized form, given in (7.226), (7.231), (7.237), and (7.242), define the new independent variable z to be a coefficient of either $w(x)$, $w'(x)$, or $w''(x)$ raised to the power $\frac{1}{2}$, 1 or 2.

7.6.2 Solution of Second Order Homogeneous Difference Equations with Linear Coefficients Through Transformation of Dependent and Independent Variables

In this section we analyze the difference equations denoted earlier by $(1, 1, 0)$, $(0, 1, 1)$, $(1, 0, 0)$, $(0, 0, 1)$ and $(0, 1, 0)$. However, rather than deriving the solution in the form of an integral, we present an approach in which the dependent and independent variables of the difference equation are transformed so that the resulting equation is recognized to have a form whose solution is a well-known special function of mathematical physics. This approach resembles that given in [13, Sect. 6.2, pp. 249–252], for differential equations of second order, where it is shown that the transformed equation has solutions which are either a confluent hypergeometric function or a Bessel function (which may be expressed in terms of confluent hypergeometric functions), the specific form of the solution depending on the coefficients.

Referring as before to the difference equation (Eq. (7.153))

$$(c_2 + d_2(x + 1))y(x + 1) + (c_1 + d_1 x)y(x) + (c_0 + d_0(x - 1))y(x - 1) = 0,$$
$$\tag{7.247}$$

we first consider the cases $(1, 1, 0)$ (for which $d_0 \neq 0, d_1 \neq 0, d_2 = 0$), and $(0, 1, 1)$ (for which $d_0 = 0, d_1 \neq 0, d_2 \neq 0$). As noted earlier, the solution to the equations defined by $(1, 1, 0)$ and $(0, 1, 1)$ are related simply by a multiplicative factor. Thus, if we start with $(1, 1, 0)$, namely

$$c_2 y(x + 1) + (c_1 + d_1 x)y(x) + (c_0 + d_0(x - 1))y(x - 1) = 0 \tag{7.248}$$

then, writing this as

$$c_2 y(x + 1) + (c_1 + d_1 x)y(x) + d_0 \left(\frac{c_0}{d_0} + x - 1 \right) y(x - 1) = 0 \tag{7.249}$$

and defining

$$y(x) = \Gamma \left(\frac{c_0}{d_0} + x \right) w(x) \tag{7.250}$$

we obtain the difference equation for $w(x)$

$$c_2 \left(\frac{c_0}{d_0} + x \right) w(x + 1) + (c_1 + d_1 x)w(x) + d_0 w(x - 1) = 0 \tag{7.251}$$

which is denoted by $(0, 1, 1)$. In similar fashion, if we start with a difference equation denoted by $(0, 1, 1)$, namely

$$(c_2 + d_2(x + 1))y(x + 1) + (c_1 + d_1 x)y(x) + c_0 y(x - 1) = 0 \tag{7.252}$$

then, writing this as

$$d_2 \left(\frac{c_2}{d_2} + x + 1 \right) y(x + 1) + (c_1 + d_1 x)y(x) + c_0 y(x - 1) = 0 \tag{7.253}$$

and defining

$$y(x) = \frac{w(x)}{\Gamma \left(\frac{c_2}{d_2} + x + 1 \right)} \tag{7.254}$$

we obtain the difference equation for $w(x)$

$$d_2 w(x + 1) + (c_1 + d_1 x)w(x) + c_0 \left(\frac{c_2}{d_2} + x \right) w(x - 1) = 0 \tag{7.255}$$

which is denoted by $(1, 1, 0)$.

We consider first the difference equation denoted by $(1, 1, 0)$, i.e.,

$$c_2 y(x+1) + (c_1 + d_1 x)y(x) + (c_0 + d_0(x-1))y(x-1) = 0 \qquad (7.256)$$

and note from (8.29) that both the confluent hypergeometric functions $M(a; c; z) = \frac{\Gamma(c-a)}{\Gamma(c)} {}_1F_1(a; c; z)$ and $U(a; c; z)$ obey a difference equation denoted by $(1, 1, 0)$, namely,

$$z F(a; c+1; z) + (1 - z - c)F(a; c; z) + (-a + c - 1)F(a; c-1; z) = 0. \qquad (7.257)$$

In order to show that the difference equation (7.256) has solutions of the form of $M(a; c; z) = \frac{\Gamma(c-a)}{\Gamma(c)} {}_1F_1(a; c; z)$ and $U(a; c; z)$, we define $\overline{F}(a; c; z) = e^{i\pi c} F(a; c; z)$ and replace c by $c + x$, giving

$$z\overline{F}(a; c+x+1; z) + (z - 1 + c + x)\overline{F}(a; c+x; z) + (-a+c+x-1)\overline{F}(a; c+x-1; z) = 0 \qquad (7.258)$$

We then define

$$y(x) = \lambda^x w(x) \qquad (7.259)$$

giving the difference equation for $w(x)$:

$$\frac{\lambda^2 c_2}{d_0} w(x+1) + \frac{\lambda}{d_0} w(x) + \left(\frac{c_0}{d_0} + x - 1\right) w(x-1) = 0 \qquad (7.260)$$

To equate the difference equations (7.260) and (7.258) for $w(x)$ and $\overline{F}(a; c+x; z)$ we define

$$\lambda = \frac{d_0}{d_1}$$

$$c - a = \frac{c_0}{d_0}$$

$$z = \frac{\lambda^2 c_2}{d_0}$$

$$c + z - 1 = \frac{\lambda c_1}{d_0} \qquad (7.261)$$

from which $z = \frac{c_2 d_0}{d_1^2}$, $c = \frac{c_1}{d_1} + 1 - \frac{c_2 d_0}{d_1^2}$, $a = \frac{c_1}{d_1} + 1 - \frac{c_2 d_0}{d_1^2} - \frac{c_0}{d_0}$.

We then have,[6] provided $a \neq 0, -1, -2, \ldots$, two linearly independent solutions of (7.256),

---

[6]If $a = 0, -1, -2\ldots$, then from [36, Sect. 13.2(i), Eq. 13.2.4], $\frac{\Gamma(c-a+x)}{\Gamma(c+x)} {}_1F_1(a; c+x; z) = (-1)^a U(a; c+x; z)$; hence $y_1(x) = (-1)^a y_2(x)$.

$$y_1(x) = \lambda^x e^{i\pi x} \frac{\Gamma(c-a+x)}{\Gamma(c+x)} {}_1F_1(a; c+x; z)$$

$$y_2(x) = \lambda^x e^{i\pi x} U(a; c+x; z) \tag{7.262}$$

The Casoratian of $y_1(x)$ and $y_2(x)$ may be derived as shown earlier, using the raising operators to relate the Casoratian to the Wronskian. From [36, Sect. 13.3(ii), Eqs. 13.3.20 and 13.3.27], we have

$$\frac{d}{dz}\left(e^{-z} {}_1F_1(a; c+x; z)\right) = -\frac{(c-a+x)}{(c+x)} e^{-z} {}_1F_1(a; c+x+1; z)$$

$$\frac{d}{dz}\left(e^{-z} U(a; c+x; z)\right) = -e^{-z} U(a; c+x+1; z) \tag{7.263}$$

from which

$$\begin{aligned}
\mathcal{C}(y_1(x), y_2(x)) &= y_1(x)y_2(x+1) - y_1(x+1)y_2(x) \\
&= \lambda^{2x+1} e^{2i\pi x} \frac{\Gamma(c-a+x)}{\Gamma(c+x)} \\
&\quad \times \left({}_1F_1(a; c+x; z)e^z \frac{d}{dz}\left(e^{-z}U(a; c+x; z)\right) \right. \\
&\qquad \left. -U(a; c+x; z)e^z \frac{d}{dz}\left(e^{-z} {}_1F_1(a; c+x; z)\right) \right) \\
&= \lambda^{2x+1} e^{2i\pi x} \frac{\Gamma(c-a+x)}{\Gamma(c+x)} \\
&\quad \times \left({}_1F_1(a; c+x; z)\frac{d}{dz}U(a; c+x; z) - \frac{d}{dz} {}_1F_1(a; c+x; z)U(a; c+x; z) \right) \\
&= \lambda^{2x+1} e^{2i\pi x} \frac{\Gamma(c-a+x)}{\Gamma(c+x)} \mathcal{W}\left({}_1F_1(a; c+x; z), U(a; c+x; z)\right) \\
&= -\lambda^{2x+1} e^{2i\pi x} \frac{\Gamma(c-a+x)}{\Gamma(a)} z^{-c-x} e^z \tag{7.264}
\end{aligned}$$

from [36, Sect. 13.2(vi), Eq. 13.2.34].

We next consider the cases $(1, 0, 0)$ (for which $d_0 \neq 0, d_1 = 0, d_2 = 0$), and $(0, 0, 1)$ (for which $d_0 = 0, d_1 = 0, d_2 \neq 0$). As noted earlier, the solution to the equations defined by $(1, 0, 0)$ and $(0, 0, 1)$ are related simply by a multiplicative factor. Thus, if we start with $(1, 0, 0)$, namely

$$c_2 y(x+1) + c_1 y(x) + (c_0 + d_0(x-1))y(x-1) = 0 \tag{7.265}$$

then, writing this as

$$c_2 y(x+1) + c_1 y(x) + d_0 \left(\frac{c_0}{d_0} + x - 1\right) y(x-1) = 0 \tag{7.266}$$

and defining

$$y(x) = \Gamma\left(\frac{c_0}{d_0} + x\right)w(x) \tag{7.267}$$

we obtain the difference equation for $w(x)$

$$c_2\left(\frac{c_0}{d_0} + x\right)w(x+1) + c_1 w(x) + d_0 w(x-1) = 0 \tag{7.268}$$

which is denoted by $(0, 0, 1)$. In similar fashion, if we start with a difference equation denoted by $(0, 0, 1)$, namely

$$(c_2 + d_2(x+1))y(x+1) + c_1 y(x) + c_0 y(x-1) = 0 \tag{7.269}$$

then writing this as

$$d_2\left(\frac{c_2}{d_2} + x + 1\right) + c_1 y(x) + c_0 y(x-1) = 0 \tag{7.270}$$

and defining

$$y(x) = \frac{w(x)}{\Gamma(\frac{c_2}{d_2} + x + 1)} \tag{7.271}$$

we obtain the difference equation for $w(x)$

$$d_2 w(x+1) + c_1 w(x) + c_0\left(\frac{c_2}{d_2} + x\right)w(x-1) = 0 \tag{7.272}$$

which is denoted by $(1, 0, 0)$.

We consider next the difference equation denoted by $(1, 0, 0)$, i.e.,

$$c_2 y(x+1) + c_1 y(x) + (c_0 + d_0(x-1))y(x-1) = 0 \tag{7.273}$$

and note from (8.36) and (8.38) that both of the parabolic cylinder functions $W(a, z) = e^{i\pi a}\Gamma(a + \frac{1}{2})U(a, z)$ and $V(a, z)$ obey a difference equation denoted by $(1, 0, 0)$, namely

$$F(a+1, z) - zF(a, z) - (a - \tfrac{1}{2})F(a-1, z) = 0. \tag{7.274}$$

In order to show that the difference equation (7.273) has solutions of the form $W(a, z) = e^{i\pi a}\Gamma(a + \frac{1}{2})U(a, z)$ and $V(a, z)$, we replace a by $a + x$ and define

$$y(x) = \lambda^x w(x) \tag{7.275}$$

giving the difference equation for $w(x)$:

$$w(x+1) + \frac{c_1}{c_2\lambda}w(x) + \frac{d_0}{c_2\lambda^2}\left(\frac{c_0}{d_0} + x - 1\right)w(x-1) = 0 \tag{7.276}$$

To equate the difference equations (7.276) and (7.274) for $w(x)$ and $F(a+x, z)$ we define[7]

$$\lambda^2 = -\frac{d_0}{c_2}$$

$$a = -\left(\frac{c_0}{c_2} + \frac{1}{2}\right)$$

$$z = -\frac{c_1}{c_2\lambda} \tag{7.277}$$

We then have, provided $a + x \neq -\frac{1}{2}, -\frac{3}{2}, -\frac{5}{2}, \ldots,$ two linearly independent solutions of (7.273),

$$y_1(x) = \lambda^x e^{i\pi x}\Gamma(a + x + \tfrac{1}{2})U(a + x, z)$$
$$y_2(x) = \lambda^x V(a + x, z) \tag{7.278}$$

The Casoratian of $y_1(x)$ and $y_2(x)$ may be derived as shown earlier using the raising operators to relate the Casoratian to the Wronskian. From [36, Sect. 12.8(ii), Eqs. 12.8.9 and 12.8.11] we have

$$\frac{d}{dz}\left(e^{\frac{1}{4}z^2}U(a+x, z)\right) = -(a + x + \tfrac{1}{2})e^{\frac{1}{4}z^2}U(a + x + 1, z)$$
$$\frac{d}{dz}\left(e^{\frac{1}{4}z^2}V(a+x, z)\right) = e^{\frac{1}{4}z^2}V(a + x + 1, z) \tag{7.279}$$

from which

$$\mathcal{C}(y_1(x), y_2(x)) = y_1(x)y_2(x+1) - y_1(x+1)y_2(x)$$
$$= \lambda^{2x+1}e^{i\pi x}\Gamma(a + x + \tfrac{1}{2})e^{-\frac{1}{4}z^2}$$
$$\times \left(U(a+x)\frac{d}{dz}\left(e^{\frac{1}{4}z^2}V(a+x, z)\right) - V(a+x, z)\frac{d}{dz}\left(e^{\frac{1}{4}z^2}U(a+x, z)\right)\right)$$
$$= \lambda^{2x+1}e^{i\pi x}\Gamma(a + x + \tfrac{1}{2})$$
$$\times \left(U(a+x)\frac{d}{dz}V(a+x, z) - V(a+x, z)\frac{d}{dz}U(a+x, z)\right)$$
$$= \lambda^{2x+1}e^{i\pi x}\Gamma(a + x + \tfrac{1}{2})\mathcal{W}(U(a+x), V(a+x))$$
$$= \lambda^{2x+1}e^{i\pi x}\Gamma(a + x + \tfrac{1}{2})\sqrt{\frac{2}{\pi}} \tag{7.280}$$

from [36, Sect. 12.8(ii), Eq. 12.2.10].

---

[7]If $\frac{d_0}{c_2} < 0$ we can choose $\lambda = -\text{sgn}\left(\frac{c_1}{c_2}\right)\sqrt{-\frac{d_0}{c_2}}$ so that $z > 0$.

We consider finally the difference equation denoted by $(0, 1, 0)$, i.e.,

$$c_2 y(x+1) + (c_{1+d_1 x}) y(x) + c_0 y(x-1) = 0 \qquad (7.281)$$

Here, there are two cases to consider: (a) c_0 and c_2 have the same sign (hence $c_0 c_2 > 0$), and (b) c_0 and c_2 have the opposite sign (hence $c_0 c_2 < 0$). If $c_0 c_2 > 0$, then writing $y(x) = \lambda^x w(x)$, the difference equation for $w(x)$ is

$$c_2 \lambda^2 w(x+1) + \lambda(c_1 + d_1 x) w(x) + c_0 w(x-1) = 0. \qquad (7.282)$$

Choosing $\lambda = -\mathrm{sgn}(d_1) \sqrt{\frac{c_0}{c_2}}$. We then have

$$w(x+1) - \frac{|d_1|}{\sqrt{c_0 c_2}} \left(\frac{c_1}{d_1} + x \right) w(x) + w(x-1) = 0. \qquad (7.283)$$

Then from (8.32) we have the four linearly independent solutions $J_\nu(z)$, $Y_\nu(z)$, $H_\nu^{(1)}(z)$, $H_\nu^{(2)}(z)$, in which

$$z = \frac{2\sqrt{c_1 c_2}}{|d_1|} \qquad \nu = \frac{c_1}{d_1} + x \qquad (7.284)$$

As before, the Casoratians can be obtained using the raising operators. However, in this case they are, apart from a sign, equal to the Wronskian, and are given in [36, Sect. 10.5, Eqs. 10.5.2–10.5.5]:

$$\mathcal{C}\left(J_\nu(z), Y_\nu(z)\right) = -\frac{2}{\pi z}$$

$$\mathcal{C}\left(J_\nu(z), H_\nu^{(1)}(z)\right) = -\frac{2i}{\pi z}$$

$$\mathcal{C}\left(J_\nu(z), H_\nu^{(2)}(z)\right) = \frac{2i}{\pi z}$$

$$\mathcal{C}\left(H_\nu^{(1)}(z), H_\nu^{(2)}(z)\right) = \frac{4i}{\pi z} \qquad (7.285)$$

Chapter 8
Dictionary of Difference Equations with Polynomial Coefficients

In this chapter we list the difference equations for some of the classical functions and polynomials of mathematical physics. These difference equations have coefficients which are polynomials in the argument of the difference equation, their degree ranging from zero (i.e., constant coefficients) for the Tchebichef polynomials $T_n(x)$ and $U_n(x)$ to three for the Jacobi polynomials $P_n^{(\alpha,\beta)}(x)$ and $Q_n^{(\alpha,\beta)}(x)$. We list only those difference equations in which only one parameter varies (note, for example, Eq. (8.16), in which only ν varies, or Eq. (8.17), in which only μ varies). A full investigation of the relations between contiguous functions, which involve two of the parameters (given, e.g., in [13, 36]) leads to the subject of partial difference equations, which we leave to future work. We note, however, that for a function with two parameters (e.g., $P_\nu^\mu(x)$), one can obtain a three-term equation connecting the function at any three distinct points in the two-dimensional grid $(a + n, b + m)$: $AF(a + n_1, b + m_1) + BF(a + n_2, b + m_2) + CF(a + n_3, b + m_3) = 0$ directly from the four contiguous relations for the nine contiguous values of the function, for which $n, m = 0, \pm 1$:

$$AF(a, b) + BF(a + 1, b) + CF(a, b + 1) = 0$$
$$AF(a, b) + BF(a - 1, b) + CF(a, b + 1) = 0$$
$$AF(a, b) + BF(a + 1, b) + CF(a, b - 1) = 0$$
$$AF(a, b) + BF(a - 1, b) + CF(a, b - 1) = 0,$$

$$(8.1)$$

Here, (n_1, m_1), (n_2, m_2) and (n_3, m_3) are distinct pairs of integers, $n_i, m_i = 0, \pm 1, \pm 2, \ldots$, and A, B and C are functions of a and b. In particular, the two relations involving only one parameter,

$$AF(a, b) + BF(a, b + 1) + CF(a, b - 1) = 0 \qquad (8.2)$$

© Springer International Publishing Switzerland 2016
L.C. Maximon, *Differential and Difference Equations*,
DOI 10.1007/978-3-319-29736-1_8

and

$$AF(a, b) + BF(a + 1, b) + CF(a - 1, b) = 0 \qquad (8.3)$$

can be derived directly from the contiguous relations given above.

Difference Equations with Constant Coefficients

The Tchebichef polynomials

$$T_n(\cos \theta) = \cos n\theta$$
$$U_n(\cos \theta) = \frac{\sin(n + 1)\theta}{\sin \theta} \qquad (8.4)$$

obey the same difference equation:

$$T_{n+1}(x) - 2xT_n(x) + T_{n-1}(x) = 0 \qquad (8.5)$$

$$U_{n+1}(x) - 2xU_n(x) + U_{n-1}(x) = 0 \qquad (8.6)$$

Difference Equations with Linear Coefficients

The parameters c_i and d_i, $(i = 0, 1, 2)$, in this section refer to Eq. (7.153). See also the discussion surrounding (7.163).

(i) Orthogonal polynomials

(i1) Legendre polynomials

$$(n + 1)P_{n+1}(z) - 2z(n + \tfrac{1}{2})P_n(z) + nP_{n-1}(z) = 0$$
$$c_2 = 0, \quad c_1 = -z, \quad c_0 = 1$$
$$d_2 = 1, \quad d_1 = -2z, \quad d_0 = 1, \quad d_1^2 - 4d_0d_2 = 4(z^2 - 1) \qquad (8.7)$$
$$\left(\frac{c_0}{d_0} - 2\frac{c_1}{d_1} + \frac{c_2}{d_2} \right) = 0$$

(i2) Gegenbauer polynomials, $C_n^{\nu}(z)$. The Gegenbauer polynomials, also called ultraspherical polynomials, can be expressed in terms of Legendre functions: $C_n^{\nu}(z) = 2^{-\nu + \frac{1}{2}} \pi^{\frac{1}{2}} \frac{\Gamma(n+2\nu)}{\Gamma(\nu)n!} (z^2 - 1)^{\frac{1}{4} - \frac{1}{2}\nu} P_{n+\nu-\frac{1}{2}}^{\frac{1}{2}-\nu}(z)$ as well as in terms of Jacobi polynomials: $C_n^{\nu}(z) = \frac{\Gamma(\nu+\frac{1}{2})}{\Gamma(n+\nu+\frac{1}{2})} \frac{\Gamma(n+2\nu)}{\Gamma(2\nu)} P_n^{(\nu-\frac{1}{2}, \nu-\frac{1}{2})}(z)$.

$$(n + 1)C_{n+1}^{\nu}(z) - 2z(n + \nu)C_n^{\nu}(z) + (n - 1 + 2\nu)C_{n-1}^{\nu}(z) = 0$$
$$c_2 = 0, \quad c_1 = -2\nu z, \quad c_0 = 2\nu$$
$$d_2 = 1, \quad d_1 = -2z, \quad d_0 = 1, \quad d_1^2 - 4d_0d_2 = 4(z^2 - 1) \qquad (8.8)$$
$$\left(\frac{c_0}{d_0} - 2\frac{c_1}{d_1} + \frac{c_2}{d_2} \right) = 0$$

(i3) Laguerre polynomials

$$(n+1)L_{n+1}^{\alpha}(x) - (2n+\alpha+1-x)L_n^{\alpha}(x) + (n+\alpha)L_{n-1}^{\alpha}(x) = 0$$
$$c_2 = 0, \quad c_1 = -(\alpha+1-x), \quad c_0 = \alpha+1$$
$$d_2 = 1, \quad d_1 = -2, \quad d_0 = 1, \quad d_1^2 - 4d_0d_2 = 0 \tag{8.9}$$
$$\left(\frac{c_0}{d_0} - 2\frac{c_1}{d_1} + \frac{c_2}{d_2}\right) = x$$

(i4) Laguerre polynomials

$$xL_n^{\alpha+1}(x) - (\alpha+x)L_n^{\alpha}(x) + (\alpha+n)L_n^{\alpha-1}(x) = 0$$
$$c_2 = x, \quad c_1 = -x, \quad c_0 = n+1 \tag{8.10}$$
$$d_2 = 0, \quad d_1 = -1, \quad d_0 = 1, \quad d_1^2 - 4d_0d_2 = 1$$

(i5) Hermite polynomials

$$H_{n+1}(x) - 2xH_n(x) + 2nH_{n-1}(x) = 0$$
$$c_2 = 1, \quad c_1 = -2x, \quad c_0 = 2 \tag{8.11}$$
$$d_2 = 0, \quad d_1 = 0, \quad d_0 = 2, \quad d_1^2 - 4d_0d_2 = 0$$

(i6) Hypergeometric polynomial $F_n(z) \equiv {}_2F_1(-n, b; c; z)$

$$(n+c)F_{n+1}(z) - [c+2n-(b+n)z]F_n(z) + (1-z)nF_{n-1}(z) = 0$$
$$c_2 = c-1, \quad c_1 = bz-c, \quad c_0 = 1-z$$
$$d_2 = 1, \quad d_1 = z-2, \quad d_0 = 1-z, \quad d_1^2 - 4d_0d_2 = z^2 \tag{8.12}$$
$$\left(\frac{c_0}{d_0} - 2\frac{c_1}{d_1} + \frac{c_2}{d_2}\right) = \frac{z(c-2b)}{z-2}$$

(i7) Associated Legendre polynomial $P_n^{-\alpha}(z) = \left(\frac{z-1}{z+1}\right)^{\frac{1}{2}\alpha} {}_2F_1(-n, n+1; 1+\alpha; \frac{1-z}{2})/\Gamma(1+\alpha)$.

The polynomial $P_n^{-\alpha}(z)$ defined here can be expressed in terms of the Jacobi polynomial $P_n^{(\alpha,\beta)}(z)$ with $\beta = -\alpha$: $P_n^{(\alpha,-\alpha)}(z) = \frac{\Gamma(1+n+\alpha)}{n!}\left(\frac{z+1}{z-1}\right)^{\frac{1}{2}\alpha} P_n^{-\alpha}(z)$.

$$(\alpha+n+1)P_{n+1}^{-\alpha}(z) - 2z(n+\tfrac{1}{2})P_n^{-\alpha}(z) + (n-\alpha)P_{n-1}^{-\alpha}(z) = 0$$
$$c_2 = \alpha, \quad c_1 = -z, \quad c_0 = 1-\alpha$$
$$d_2 = 1, \quad d_1 = -2z, \quad d_0 = 1, \quad d_1^2 - 4d_0d_2 = 4(z^2-1) \tag{8.13}$$
$$\left(\frac{c_0}{d_0} - 2\frac{c_1}{d_1} + \frac{c_2}{d_2}\right) = 0$$

(i8) Jacobi polynomial $P_n^{(\alpha,\beta)}(x)$. A second solution of the differential equation obeyed by the function $P_n^{(\alpha,\beta)}(x)$ is $Q_n^{(\alpha,\beta)}(x)$, which is not a polynomial but satisfies the same recurrence relations as $P_n^{(\alpha,\beta)}(x)$ and is a second, linearly independent, solution of the difference equation satisfied by $P_n^{(\alpha,\beta)}(x)$:

$$(n+\alpha+\beta+1)(1-x)P_n^{(\alpha+1,\beta)}(x) - (2\alpha + (2n+\alpha+\beta+1)(1-x))P_n^{(\alpha,\beta)}(x)$$
$$+ 2(n+\alpha)P_n^{(\alpha-1,\beta)}(x) = 0$$
$$c_2 = (n+\beta)(1-x), \quad c_1 = -(2n+\beta+1)(1-x), \quad c_0 = 2(n+1)$$
$$d_2 = (1-x), \quad d_1 = -(3-x), \quad d_0 = 2, \quad d_1^2 - 4d_0 d_2 = (1+x)^2$$
$$\left(\frac{c_0}{d_0} - 2\frac{c_1}{d_1} + \frac{c_2}{d_2} \right) = \frac{(2n+\beta+1)(1+x)}{3-x} \tag{8.14}$$

(i9) Jacobi polynomial $P_n^{(\alpha,\beta)}(x)$. A second solution of the differential equation obeyed by the function $P_n^{(\alpha,\beta)}(x)$ is $Q_n^{(\alpha,\beta)}(x)$, which is not a polynomial but satisfies the same recurrence relations as $P_n^{(\alpha,\beta)}(x)$ and is a second, linearly independent, solution of the difference equation satisfied by $P_n^{(\alpha,\beta)}(x)$:

$$(n+\alpha+\beta+1)(1+x)P_n^{(\alpha,\beta+1)}(x) - (2\beta + (2n+\alpha+\beta+1)(1+x))P_n^{(\alpha,\beta)}(x)$$
$$+ 2(n+\beta)P_n^{(\alpha,\beta-1)}(x) = 0$$
$$c_2 = (n+\alpha)(1+x), \quad c_1 = -(2n+\alpha+1)(1+x), \quad c_0 = 2(n+1)$$
$$d_2 = (1+x), \quad d_1 = -(3+x), \quad d_0 = 2, \quad d_1^2 - 4d_0 d_2 = (1-x)^2$$
$$\left(\frac{c_0}{d_0} - 2\frac{c_1}{d_1} + \frac{c_2}{d_2} \right) = \frac{(2n+\alpha+1)(1-x)}{3+x} \tag{8.15}$$

(ii) Classical functions

(ii1) Legendre functions, $P_\nu^\mu(z)$. A second, linearly independent, solution of the differential equation obeyed by the function $P_\nu^\mu(z)$ is $Q_\nu^\mu(z)$, which satisfies the same recurrence relations as $P_\nu^\mu(z)$:

$$(\nu - \mu + 1)P_{\nu+1}^\mu(z) - 2z(\nu + \tfrac{1}{2})P_\nu^\mu(z) + (\nu + \mu)P_{\nu-1}^\mu(z) = 0$$
$$c_2 = -\mu, \quad c_1 = -z, \quad c_0 = \mu + 1$$
$$d_2 = 1, \quad d_1 = -2z, \quad d_0 = 1, \quad d_1^2 - 4d_0 d_2 = 4(z^2 - 1) \tag{8.16}$$
$$\left(\frac{c_0}{d_0} - 2\frac{c_1}{d_1} + \frac{c_2}{d_2} \right) = 0$$

(ii2) Legendre functions $\mathcal{P}_\nu^\mu(z) = P_\nu^\mu(z)/\Gamma(\mu - \nu)$. The function $\mathfrak{Q}_\nu^\mu(z) = Q_\nu^\mu(z)/\Gamma(\mu - \nu)$ satisfies the same recurrence relations as $\mathcal{P}_\nu^\mu(z)$:

$$(\mu - \nu)\mathcal{P}_\nu^{\mu+1}(z) + 2\mu z(z^2 - 1)^{-\frac{1}{2}}\mathcal{P}_\nu^\mu(z) + (\mu + \nu)\mathcal{P}_\nu^{\mu-1}(z)$$

$$c_2 = -(\nu + 1), \quad c_1 = 0, \quad c_0 = \nu + 1$$

$$d_2 = 1, \quad d_1 = 2z(z^2 - 1)^{-\frac{1}{2}}, \quad d_0 = 1, \quad d_1^2 - 4d_0 d_2 = \frac{4}{z^2 - 1} \tag{8.17}$$

$$\left(\frac{c_0}{d_0} - 2\frac{c_1}{d_1} + \frac{c_2}{d_2}\right) = 0$$

(ii3) Gegenbauer (ultraspherical) functions, $C_\alpha^\nu(z)$. A second, linearly independent, solution of the differential equation obeyed by the function $C_\alpha^\nu(z)$ is $D_\alpha^\nu(z)$, which satisfies the same recurrence relation in α as $C_\alpha^\nu(z)$:

$$(\alpha + 1)C_{\alpha+1}^\nu(z) - 2z(\nu + \alpha)C_\alpha^\nu(z) + (2\nu + \alpha - 1)C_{\alpha-1}^\nu(z) = 0$$

$$c_2 = 0, \quad c_1 = -2z\nu, \quad c_0 = 2\nu$$

$$d_2 = 1, \quad d_1 = -2z, \quad d_0 = 1, \quad d_1^2 - 4d_0 d_2 = 4(z^2 - 1) \tag{8.18}$$

$$\left(\frac{c_0}{d_0} - 2\frac{c_1}{d_1} + \frac{c_2}{d_2}\right) = 0$$

(ii4) Gegenbauer (ultraspherical) function, $\mathcal{C}_\alpha^\nu(z) = \dfrac{\Gamma(\nu)}{\Gamma(\nu + \frac{1}{2} + \frac{1}{2}\alpha)}C_\alpha^\nu(z)$

$$(\nu + \tfrac{1}{2} + \tfrac{1}{2}\alpha)(z^2 - 1)\mathcal{C}_\alpha^{\nu+1}(z) + [(2\nu + \alpha - \tfrac{1}{2}) - (\nu + \alpha)z^2]\mathcal{C}_\alpha^\nu(z)$$

$$- (\nu - 1 + \tfrac{1}{2}\alpha)\mathcal{C}_\alpha^{\nu-1}(z) = 0$$

$$c_2 = (-\tfrac{1}{2} + \tfrac{1}{2}\alpha)(z^2 - 1), \quad c_1 = \alpha - \tfrac{1}{2} - \alpha z^2, \quad c_0 = -\tfrac{1}{2}\alpha$$

$$d_2 = z^2 - 1, \quad d_1 = 2 - z^2, \quad d_0 = -1 \quad d_1^2 - 4d_0 d_2 = z^4 \tag{8.19}$$

$$\left(\frac{c_0}{d_0} - 2\frac{c_1}{d_1} + \frac{c_2}{d_2}\right) = \frac{z^2\left(\alpha + \frac{1}{2}\right)}{2 - z^2}$$

(ii5) Gegenbauer (ultraspherical) function, $\mathcal{C}_\alpha^\nu(z) = \dfrac{\Gamma(\nu)}{\Gamma(\nu + \frac{1}{2}\alpha)}C_\alpha^\nu(z)$

$$(\nu + \tfrac{1}{2}\alpha)(z^2 - 1)\mathcal{C}_\alpha^{\nu+1}(z) + [(2\nu + \alpha - \tfrac{1}{2}) - (\nu + \alpha)z^2]\mathcal{C}_\alpha^\nu(z)$$

$$- (\nu - \tfrac{1}{2} + \tfrac{1}{2}\alpha)\mathcal{C}_\alpha^{\nu-1}(z) = 0$$

$$c_2 = (-1 + \tfrac{1}{2}\alpha)(z^2 - 1), \quad c_1 = \alpha - \tfrac{1}{2} - \alpha z^2, \quad c_0 = -\tfrac{1}{2} - \tfrac{1}{2}\alpha \tag{8.20}$$

$$d_2 = z^2 - 1, \quad d_1 = 2 - z^2, \quad d_0 = -1 \quad d_1^2 - 4d_0 d_2 = z^4$$

$$\left(\frac{c_0}{d_0} - 2\frac{c_1}{d_1} + \frac{c_2}{d_2}\right) = \frac{z^2\left(\alpha + \frac{1}{2}\right)}{2 - z^2}$$

(ii6) Gegenbauer (ultraspherical) function, $\mathcal{D}_\alpha^\nu(z) = \frac{2^{-2\nu}(1-z^2)^\nu}{\Gamma(\nu)\Gamma(\nu+\frac{1}{2}+\frac{1}{2}\alpha)} D_\alpha^\nu(z)$

$$(\nu + \tfrac{1}{2} + \tfrac{1}{2}\alpha)z^2 \mathcal{D}_\alpha^{\nu+1}(z) + [(\nu+\alpha) - (2\nu+\alpha-\tfrac{1}{2})z^2]\mathcal{D}_\alpha^\nu(z)$$
$$- (\nu - 1 + \tfrac{1}{2}\alpha)(1 - z^2)\mathcal{D}_\alpha^{\nu-1}(z) = 0$$
$$c_2 = (-\tfrac{1}{2} + \tfrac{1}{2}\alpha)z^2, \quad c_1 = \alpha - (\alpha - \tfrac{1}{2})z^2, \quad c_0 = \tfrac{1}{2}\alpha(z^2 - 1)$$
$$d_2 = z^2, \quad d_1 = 1 - 2z^2, \quad d_0 = (z^2 - 1) \quad d_1^2 - 4d_0d_2 = 1 \qquad (8.21)$$
$$\left(\frac{c_0}{d_0} - 2\frac{c_1}{d_1} + \frac{c_2}{d_2}\right) = \frac{(\alpha + \tfrac{1}{2})}{2z^2 - 1}$$

(ii7) Gegenbauer (ultraspherical) function, $\mathcal{D}_\alpha^\nu(z) = \frac{2^{-2\nu}(1-z^2)^\nu}{\Gamma(\nu)\Gamma(\nu+\frac{1}{2}\alpha)} D_\alpha^\nu(z)$

$$(\nu + \tfrac{1}{2}\alpha)z^2 \mathcal{D}_\alpha^{\nu+1}(z) + [(\nu+\alpha) - (2\nu+\alpha-\tfrac{1}{2})z^2]\mathcal{D}_\alpha^\nu(z)$$
$$- (\nu - \tfrac{1}{2} + \tfrac{1}{2}\alpha)(1 - z^2)\mathcal{D}_\alpha^{\nu-1}(z) = 0$$
$$c_2 = (-1 + \tfrac{1}{2}\alpha)z^2, \quad c_1 = \alpha - (\alpha - \tfrac{1}{2})z^2, \quad c_0 = (\tfrac{1}{2} + \tfrac{1}{2}\alpha)(z^2 - 1)$$
$$d_2 = z^2, \quad d_1 = 1 - 2z^2, \quad d_0 = (z^2 - 1) \quad d_1^2 - 4d_0d_2 = 1 \qquad (8.22)$$
$$\left(\frac{c_0}{d_0} - 2\frac{c_1}{d_1} + \frac{c_2}{d_2}\right) = \frac{(\alpha + \tfrac{1}{2})}{2z^2 - 1}$$

(ii8) Hypergeometric function $_2F_1(a, b; c; z)$

$$a(1 - z)_2F_1(a + 1, b; c; z) + [c - 2a + (a - b)z]_2F_1(a, b; c; z)$$
$$+ (a - c)_2F_1(a - 1, b; c; z) = 0$$
$$c_2 = z - 1, \quad c_1 = c - bz, \quad c_0 = 1 - c$$
$$d_2 = 1 - z, \quad d_1 = z - 2, \quad d_0 = 1, \quad d_1^2 - 4d_0d_2 = z^2 \qquad (8.23)$$
$$\left(\frac{c_0}{d_0} - 2\frac{c_1}{d_1} + \frac{c_2}{d_2}\right) = \frac{z(c - 2b)}{2 - z}$$

(ii9) Hypergeometric function $_2F_1(a, b; c; z)$

$$b(1 - z)_2F_1(a, b + 1; c; z) + [c - 2b + (b - a)z]_2F_1(a, b; c; z)$$
$$+ (b - c)_2F_1(a, b - 1; c; z) = 0$$
$$c_2 = z - 1, \quad c_1 = c - az, \quad c_0 = 1 - c$$
$$d_2 = 1 - z, \quad d_1 = z - 2, \quad d_0 = 1, \quad d_1^2 - 4d_0d_2 = z^2 \qquad (8.24)$$
$$\left(\frac{c_0}{d_0} - 2\frac{c_1}{d_1} + \frac{c_2}{d_2}\right) = \frac{z(c - 2a)}{2 - z}$$

(ii10) Hypergeometric function[1] $_2\mathcal{F}_1(a, b; c; z) = \frac{\Gamma(c-b)}{\Gamma(c)} {}_2F_1(a, b; c; z)$

$$(c - a)z\,_2\mathcal{F}_1(a, b; c + 1; z) + (c - 1 - (2c - a - b - 1)z)\,_2\mathcal{F}_1(a, b; c; z)$$
$$- (c - b - 1)(1 - z)\,_2\mathcal{F}_1(a, b; c - 1; z) = 0$$
$$c_2 = -(a + 1)z, \quad c_1 = (a + b + 1)z - 1, \quad c_0 = b(1 - z) \qquad (8.25)$$
$$d_2 = z, \quad d_1 = 1 - 2z, \quad d_0 = -(1 - z), \quad d_1^2 - 4d_0d_2 = 1$$
$$\left(\frac{c_0}{d_0} - 2\frac{c_1}{d_1} + \frac{c_2}{d_2}\right) = \frac{1 - a - b}{1 - 2z}$$

(ii11)   Confluent   hypergeometric   function $F(a; c; z) = {}_1F_1(a; c; z)$ or
$\Gamma(a + 1 - c)U(a; c; z)$

$$a\,F(a + 1; c; z) - (2a - c + z)\,F(a; c; z) + (a - c)\,F(a - 1; c; z) = 0$$
$$c_2 = -1, \quad c_1 = c - z, \quad c_0 = 1 - c$$
$$d_2 = 1, \quad d_1 = -2, \quad d_0 = 1, \quad d_1^2 - 4d_0d_2 = 0 \qquad (8.26)$$
$$\left(\frac{c_0}{d_0} - 2\frac{c_1}{d_1} + \frac{c_2}{d_2}\right) = -z$$

(ii12)   Confluent   hypergeometric   function $F(a; c; z) = \frac{\Gamma(c-a)}{\Gamma(1-a)} {}_1F_1(a; c; z)$,
$\frac{\Gamma(a)}{\Gamma(a+1-c)} {}_1F_1(a; c; z)$, or $\Gamma(a)U(a; c; z)$

$$(a + 1 - c)\,F(a + 1; c; z) - (2a - c + z)\,F(a; c; z) + (a - 1)\,F(a - 1; c; z) = 0$$
$$c_2 = -c, \quad c_1 = c - z, \quad c_0 = 0$$
$$d_2 = 1, \quad d_1 = -2, \quad d_0 = 1, \quad d_1^2 - 4d_0d_2 = 0 \qquad (8.27)$$
$$\left(\frac{c_0}{d_0} - 2\frac{c_1}{d_1} + \frac{c_2}{d_2}\right) = -z$$

(ii13)   Confluent   hypergeometric   function $F(a; c; z) = {}_1F_1(a; c; z)/\Gamma(c)$ or
$U(a; c; z)/\Gamma(c - a)$

$$z(c - a)F(a; c + 1; z) + (1 - z - c)F(a; c; z) + F(a; c - 1; z) = 0$$
$$c_2 = -z(a + 1), \quad c_1 = 1 - z, \quad c_0 = 1 \qquad (8.28)$$
$$d_2 = z, \quad d_1 = -1, \quad d_0 = 0, \quad d_1^2 - 4d_0d_2 = 1$$

---

[1] In Eqs. (8.25)–(8.31) in which functions have a coefficient of the form $\Gamma(b)$, one may replace $\Gamma(b)$ by $e^{i\pi b}/\Gamma(1 - b)$ without modifying the associated difference equation.

(ii14) Confluent hypergeometric function $F(a; c; z) = \frac{\Gamma(c-a)}{\Gamma(c)} {}_1F_1(a; c; z)$ or $U(a; c; z)$

$$zF(a; c+1; z) + (1 - z - c)F(a; c; z) + (c - 1 - a)F(a; c - 1; z) = 0$$
$$c_2 = z, \quad c_1 = 1 - z, \quad c_0 = -a \tag{8.29}$$
$$d_2 = 0, \quad d_1 = -1, \quad d_0 = 1, \quad d_1^2 - 4d_0d_2 = 1$$

(ii15) Confluent hypergeometric function $U(a; c; z)$

$$zaU(a + 1; c + 1; z) + (z - c + 1)U(a; c; z) - U(a - 1; c - 1; z) = 0$$
$$c_2 = 0, \quad c_1 = z + 1, \quad c_0 = -1 \tag{8.30}$$
$$d_2 = z, \quad d_1 = -1, \quad d_0 = 0, \quad d_1^2 - 4d_0d_2 = 1$$

(ii16) Confluent hypergeometric function $F(a; c; z) = \Gamma(a)U(a; c; z)$

$$zF(a + 1; c + 1; z) + (z - c + 1)F(a; c; z) - (a - 1)F(a - 1; c - 1; z) = 0$$
$$c_2 = z, \quad c_1 = z + 1, \quad c_0 = 0 \tag{8.31}$$
$$d_2 = 0, \quad d_1 = -1, \quad d_0 = -1, \quad d_1^2 - 4d_0d_2 = 1$$

(ii17) Bessel functions $J_\nu(z)$, $Y_\nu(z)$, $H_\nu^{(1)}(z)$, $H_\nu^{(2)}(z)$

$$J_{\nu+1}(z) - \frac{2\nu}{z} J_\nu(z) + J_{\nu-1}(z) = 0$$
$$Y_{\nu+1}(z) - \frac{2\nu}{z} Y_\nu(z) + Y_{\nu-1}(z) = 0$$
$$H_{\nu+1}^{(1)}(z) - \frac{2\nu}{z} H_\nu^{(1)}(z) + H_{\nu-1}^{(1)}(z) = 0$$
$$H_{\nu+1}^{(2)}(z) - \frac{2\nu}{z} H_\nu^{(2)}(z) + H_{\nu-1}^{(2)}(z) = 0 \tag{8.32}$$
$$c_2 = 1, \quad c_1 = 0, \quad c_0 = 1$$
$$d_2 = 0, \quad d_1 = -\frac{2}{z}, \quad d_0 = 0, \quad d_1^2 - 4d_0d_2 = \frac{4}{z^2}$$

(ii18) Bessel functions $I_\nu(z)$, $\overline{K}_\nu(z) = e^{\nu\pi i} K_\nu(z)$

$$I_{\nu+1}(z) + \frac{2\nu}{z} I_\nu(z) - I_{\nu-1}(z) = 0$$
$$\overline{K}_{\nu+1}(z) + \frac{2\nu}{z} \overline{K}_\nu(z) - \overline{K}_{\nu-1}(z) = 0 \tag{8.33}$$
$$c_2 = 1, \quad c_1 = 0, \quad c_0 = -1$$
$$d_2 = 0, \quad d_1 = \frac{2}{z}, \quad d_0 = 0, \quad d_1^2 - 4d_0d_2 = \frac{4}{z^2}$$

(ii19) Bessel function $K_\nu(z)$

$$K_{\nu+1}(z) - \frac{2\nu}{z} K_\nu(z) - K_{\nu-1}(z) = 0$$

$$c_2 = 1, \quad c_1 = 0, \quad c_0 = -1 \tag{8.34}$$

$$d_2 = 0, \quad d_1 = -\frac{2}{z}, \quad d_0 = 0, \quad d_1^2 - 4d_0 d_2 = \frac{4}{z^2}$$

(ii20) Parabolic cylinder function $D_\nu(z)$

$$D_{\nu+1}(z) - z D_\nu(z) + \nu D_{\nu-1}(z) = 0$$

$$c_2 = 1, \quad c_1 = -z, \quad c_0 = 1 \tag{8.35}$$

$$d_2 = 0, \quad d_1 = 0, \quad d_0 = 1, \quad d_1^2 - 4d_0 d_2 = 0$$

(ii21) Parabolic cylinder function $V(a, z)$

$$V(a + 1) - z V(a, z) - (a - \tfrac{1}{2}) V(a - 1, z) = 0$$

$$c_2 = 1, \quad c_1 = -z, \quad c_0 = -\tfrac{1}{2} \tag{8.36}$$

$$d_2 = 0, \quad d_1 = 0, \quad d_0 = -1, \quad d_1^2 - 4d_0 d_2 = 0$$

(ii22) Parabolic cylinder function $U(a, z)$

$$(a + \tfrac{1}{2}) U(a + 1) + z U(a, z) - U(a - 1, z) = 0$$

$$c_2 = -\tfrac{1}{2}, \quad c_1 = z, \quad c_0 = -1 \tag{8.37}$$

$$d_2 = 1, \quad d_1 = 0, \quad d_0 = 0, \quad d_1^2 - 4d_0 d_2 = 0$$

(ii23) Parabolic cylinder function $W(a, z) = e^{i\pi a} \Gamma(a + \tfrac{1}{2}) U(a, z)$

$$W(a + 1) - z W(a, z) - (a - \tfrac{1}{2}) W(a - 1, z) = 0$$

$$c_2 = 1, \quad c_1 = -z, \quad c_0 = -\tfrac{1}{2} \tag{8.38}$$

$$d_2 = 0, \quad d_1 = 0, \quad d_0 = -1, \quad d_1^2 - 4d_0 d_2 = 0$$

Difference Equations with Quadratic Coefficients

(iii1) Legendre functions, $P_\nu^\mu(z)$.
A second solution of the differential equation obeyed by the function $P_\nu^\mu(z)$ is $Q_\nu^\mu(z)$, which satisfies the same recurrence relations as $P_\nu^\mu(z)$ and is a second, linearly independent, solution of the difference equation satisfied by $P_\nu^\mu(z)$:

$$P_\nu^{\mu+1}(z) + 2\mu z (z^2 - 1)^{-\frac{1}{2}} P_\nu^\mu(z) + (\mu + \nu)(\mu - \nu - 1) P_\nu^{\mu-1}(z) = 0 \tag{8.39}$$

(iii2) Gegenbauer (ultraspherical) function, $C_\alpha^\nu(z)$

$$\nu(\nu-1)(z^2-1)C_\alpha^{\nu+1}(z) + (\nu-1)[(2\nu+\alpha-\tfrac{1}{2}) - (\nu+\alpha)z^2]C_\alpha^\nu(z)$$
$$- (\nu-\tfrac{1}{2}+\tfrac{1}{2}\alpha)(\nu-1+\tfrac{1}{2}\alpha)C_\alpha^{\nu-1}(z) = 0 \tag{8.40}$$

(iii3) Hypergeometric function $_2F_1(a, b; c; z)$

$$(c-a)(c-b)z\,_2F_1(a, b; c+1; z) + c(c-1-(2c-a-b-1)z)\,_2F_1(a, b; c+1; z)$$
$$- c(c-1)(1-z)\,_2F_1(a, b; c-1; z) = 0 \tag{8.41}$$

(iii4) Confluent hypergeometric function $_1F_1(a; c; z)$

$$(c-a)z\,_1F_1(a; c+1; z) - c(c-1+z)\,_1F_1(a; c; z) + c(c-1)\,_1F_1(a; c-1; z) \tag{8.42}$$

(iii5) Confluent hypergeometric function $U(a; c; z)$

$$a(a-c+1)U(a+1; c; z) - (2a-c+z)U(a; c; z) + U(a-1; c; z) \tag{8.43}$$

Difference Equations with Cubic Coefficients

(iv1) Jacobi polynomial $P_n^{(\alpha,\beta)}(x)$.
A second, linearly independent, solution of the differential equation obeyed by the function $P_n^{(\alpha,\beta)}(x)$ is $Q_n^{(\alpha,\beta)}(x)$, which is known as the Jacobi function of the second kind. This function is not a polynomial but satisfies the same recurrence relations as $P_n^{(\alpha,\beta)}(x)$:

$$2(n+1)(n+\alpha+\beta+1)(2n+\alpha+\beta)P_{n+1}^{(\alpha,\beta)}(x)$$
$$- (2n+\alpha+\beta+1)[(2n+\alpha+\beta)(2n+\alpha+\beta+2)x + \alpha^2 - \beta^2]P_n^{(\alpha,\beta)}(x)$$
$$+ 2(n+\alpha)(n+\beta)(2n+\alpha+\beta+2)P_{n-1}^{(\alpha,\beta)}(x) = 0 \tag{8.44}$$

Expression of Gegenbauer Functions $C_\alpha^\nu(z)$ and $D_\alpha^\nu(z)$ in terms of Legendre functions $P_\nu^\mu(z)$ and $Q_\nu^\mu(z)$.

For $C_\alpha^\nu(z)$ we have, from [13, Sect. 3.15.1, (14), p. 175], (with n replaced by an arbitrary, possibly complex α),

$$C_\alpha^\nu(z) = 2^{\nu-\frac{1}{2}}\frac{\Gamma(2\nu+\alpha)\Gamma(\nu+\frac{1}{2})}{\Gamma(2\nu)\Gamma(\alpha+1)}(z^2-1)^{\frac{1}{2}-\frac{1}{2}\nu}P_{\nu+\alpha-\frac{1}{2}}^{\frac{1}{2}-\nu}(z)$$
$$= \pi^{\frac{1}{2}}2^{-\nu+\frac{1}{2}}\frac{\Gamma(2\nu+\alpha)}{\Gamma(\nu)\Gamma(\alpha+1)}(z^2-1)^{\frac{1}{2}-\frac{1}{2}\nu}P_{\nu+\alpha-\frac{1}{2}}^{\frac{1}{2}-\nu}(z) \tag{8.45}$$

Next, from [13, Sect. 3.15.2, (32), p. 179], we have

$$D_\alpha^\nu(z) = 2^{-1-\alpha}\frac{\Gamma(\nu)\Gamma(2\nu+\alpha)}{\Gamma(\nu+\alpha+1)}{}_2F_1\left(\nu+\tfrac{1}{2}\alpha,\ \tfrac{1}{2}+\nu+\tfrac{1}{2}\alpha;\ \nu+\alpha+1;\ z^2\right)$$

(8.46)

and from [13, Sect. 3.2, (41), pp. 134–135],

$$e^{-i\mu\pi}Q_\lambda^\mu(w) = 2^{-1-\lambda}\pi^{\frac{1}{2}}\frac{\Gamma(1+\lambda+\mu)}{\Gamma(\lambda+\tfrac{3}{2})}w^{-1-\lambda-\mu}(w^2-1)^{\frac{1}{2}\mu}$$

$$\times\, {}_2F_1\left(\tfrac{1}{2}+\tfrac{1}{2}\lambda+\tfrac{1}{2}\mu,\ 1+\tfrac{1}{2}\lambda+\tfrac{1}{2}\mu;\ \lambda+\tfrac{3}{2};\ \frac{1}{w^2}\right)\qquad(8.47)$$

Setting $\lambda = \nu + \alpha - \tfrac{1}{2}$, $\mu = \nu - \tfrac{1}{2}$ and $w = z^{-1}$, we have

$$D_\alpha^\nu(z) = 2^{\nu-\frac{1}{2}}\pi^{-\frac{1}{2}}\Gamma(\nu)z^{-\nu-\alpha-\frac{1}{2}}(1-z^2)^{-\frac{1}{2}\nu+\frac{1}{4}}e^{-i(\nu-\frac{1}{2})\pi}Q_{\nu+\alpha-\frac{1}{2}}^{\nu-\frac{1}{2}}\left(\frac{1}{z}\right)\qquad(8.48)$$

Appendix A
Difference Operator

For the higher-order difference of the product of two functions we may write

$$\Delta^k(u_n v_n) = (E-1)^k(u_n v_n)$$

$$= \sum_{j=0}^{k}(-1)^{k-j}\binom{k}{j}E^j(u_n v_n) \qquad (A.1)$$

Here

$$E^j(u_n v_n) = E^j u_n E^j v_n$$

$$= (1+\Delta)^j u_n (1+\Delta)^j v_n \qquad (A.2)$$

Thus

$$\Delta^k(u_n v_n) = (-1)^k \sum_{j=0}^{k}(-1)^j\binom{k}{j}\sum_{n=0}^{j}\binom{j}{n}\Delta^n u_n \sum_{m=0}^{j}\binom{j}{m}\Delta^m v_n \qquad (A.3)$$

We now wish to sum first over j. Interchanging j and n we have

$$\sum_{j=0}^{k}\sum_{n=0}^{j} = \sum_{n=0}^{k}\sum_{j=n}^{k} \qquad (A.4)$$

from which

$$\Delta^k(u_n v_n) = (-1)^k \sum_{n=0}^{k}\sum_{j=n}^{k}(-1)^j\binom{k}{j}\binom{j}{n}\Delta^n u_n \sum_{m=0}^{j}\binom{j}{m}\Delta^m v_n \qquad (A.5)$$

© Springer International Publishing Switzerland 2016
L.C. Maximon, *Differential and Difference Equations*,
DOI 10.1007/978-3-319-29736-1

We next interchange the summations over j and m, giving

$$\sum_{j=n}^{k}\sum_{m=0}^{j} = \sum_{m=0}^{n}\sum_{j=n}^{k} + \sum_{m=n+1}^{k}\sum_{j=m}^{k} \tag{A.6}$$

We have here, for the sums over j,

$$\sum_{j=n}^{k}(-1)^j \binom{k}{j}\binom{j}{n}\binom{j}{m} \qquad 0 \le m \le n \le k \tag{A.7}$$

and

$$\sum_{j=m}^{k}(-1)^j \binom{k}{j}\binom{j}{n}\binom{j}{m} \qquad 0 \le n < m \le k \tag{A.8}$$

Thus

$$\Delta^k(u_n v_n) = (-1)^k \sum_{n=0}^{k}\sum_{m=0}^{n}\sum_{j=n}^{k}(-1)^j \binom{k}{j}\binom{j}{n}\binom{j}{m}\Delta^n u_n \Delta^m v_n$$

$$+ (-1)^k \sum_{n=0}^{k}\sum_{m=n+1}^{k}\sum_{j=m}^{k}(-1)^j \binom{k}{j}\binom{j}{n}\binom{j}{m}\Delta^n u_n \Delta^m v_n$$

$$= (-1)^k \sum_{n=0}^{k}\sum_{m=0}^{n}\Delta^n u_n \Delta^m v_n \sum_{j=n}^{k}(-1)^j \binom{k}{j}\binom{j}{n}\binom{j}{m}$$

$$+ (-1)^k \sum_{n=0}^{k}\sum_{m=n}^{k}\Delta^n u_n \Delta^m v_n \sum_{j=n}^{k}(-1)^j \binom{k}{j}\binom{j}{n}\binom{j}{m}$$

$$- (-1)^k \sum_{n=0}^{k}\sum_{j=n}^{k}(-1)^j \binom{k}{j}\binom{j}{n}\binom{j}{n}\Delta^n u_n \Delta^n v_n \tag{A.9}$$

Here

$$\sum_{n=0}^{k}\sum_{m=n}^{k} = \sum_{m=0}^{k}\sum_{n=0}^{m} \tag{A.10}$$

Thus

$$\Delta^k(u_n v_n) = (-1)^k \sum_{n=0}^{k}\sum_{m=0}^{n}\Delta^n u_n \Delta^m v_n \sum_{j=n}^{k}(-1)^j \binom{k}{j}\binom{j}{n}\binom{j}{m}$$

$$+ (-1)^k \sum_{m=0}^{k} \sum_{n=0}^{m} \Delta^n u_n \Delta^m v_n \sum_{j=m}^{k} (-1)^j \binom{k}{j} \binom{j}{n} \binom{j}{m}$$

$$- (-1)^k \sum_{n=0}^{k} \Delta^n u_n \Delta^n v_n \sum_{j=n}^{k} (-1)^j \binom{k}{j} \binom{j}{n} \binom{j}{n} \qquad \text{(A.11)}$$

Thus we need

$$\sum_{j=n}^{k} (-1)^j \binom{k}{j} \binom{j}{n} \binom{j}{m} \qquad \text{for} \quad 0 \le m \le n \le j \le k$$

$$\sum_{j=m}^{k} (-1)^j \binom{k}{j} \binom{j}{n} \binom{j}{m} \qquad \text{for} \quad 0 \le n \le m \le j \le k \qquad \text{(A.12)}$$

$$\sum_{j=n}^{k} (-1)^j \binom{k}{j} \binom{j}{n} \binom{j}{n} \qquad \text{for} \quad 0 \le n \le j \le k$$

Here

$$\sum_{j=n}^{k} (-1)^j \binom{k}{j} \binom{j}{n} \binom{j}{m} = \frac{k!}{n!m!} \sum_{j=n}^{k} (-1)^j \frac{j!}{(k-j)!(j-n)!(j-m)!} \qquad \text{(A.13)}$$

In the first sum in (A.12), for which $0 \le m \le n \le j$, we set $j' = j - n$, from which

$$\sum_{j=n}^{k} (-1)^j \frac{j!}{(k-j)!(j-n)!(j-m)!} = \sum_{j'=0}^{k-n} (-1)^{j'+n} \frac{(j'+n)!}{(k-n-j')!(j'+n-m)!j'!}$$

$$= (-1)^n \frac{n!}{(n-m)!(k-n)!} \sum_{j'=0}^{k-n} \frac{(-k+n)_{j'}(n+1)_{j'}}{j'!(n+1-m)_{j'}} \qquad \text{(A.14)}$$

where we have used

$$\frac{1}{(N-j)!} = \frac{(-1)^j(-N)_j}{N!} \qquad \text{(A.15)}$$

and

$$(N+j)! = \Gamma(N+1+j) = (N+1)_j N! \qquad \text{(A.16)}$$

We thus have, for $0 \le m \le n \le k$,

$$\sum_{j=n}^{k}(-1)^{j}\binom{k}{j}\binom{j}{n}\binom{j}{m} = (-1)^{n}\frac{k!}{m!(n-m)!(k-n)!}\ {}_2F_1(-k+n,n+1;n+1-m;1)$$

(A.17)

The well-known expression for the hypergeometric function with unit argument,

$$F(a,b;c;1) = \frac{\Gamma(c)\Gamma(c-a-b)}{\Gamma(c-a)\Gamma(c-b)}$$

(A.18)

is generally given with the condition $Re(c-a-b) > 0$. For the hypergeometric function directly above we have $c-a-b = k-n-m$ which might be a negative integer or zero, in which case the expression in (A.18) is ill-defined. However, analytic continuation of the integer values of the parameters to the complex plane make each gamma function well-behaved. Thus, since the hypergeometric polynomial is well-defined, a limit to integer arguments should exist. We thus let m be given a small imaginary component. Then, with $p = k - n$,

$$\Gamma((k-n)-m) = (p-1-m)(p-2-m)\cdots(p-p-m)\Gamma(-m) \quad \text{(A.19)}$$

is well-defined, and so is

$$\begin{aligned}
F(-k+n,n+1;n+1-m;1) &= \frac{\Gamma(n-m+1)}{\Gamma(k-m+1)}(-m)(-m+1)\cdots(-m+p-1)\\
&= \frac{\Gamma(n-m+1)}{\Gamma(k-m+1)}(-m)_p\\
&= \frac{(n-m)!}{(k-m)!}(-m)_{k-n}
\end{aligned}$$

(A.20)

Using (A.15) we then have, for $0 \le m \le n \le k$,

$$F(-k+n,n+1;n+1-m;1) = \frac{(-1)^{k-n}(n-m)!m!}{(k-m)!(m+n-k)!}$$

(A.21)

and from (A.17) and (A.21)

$$\sum_{j=n}^{k}(-1)^{j}\binom{k}{j}\binom{j}{n}\binom{j}{m} = \frac{(-1)^{k}k!}{(k-n)!(k-m)!(n+m-k)!}$$

(A.22)

Noting that the right-hand side of this equation is symmetric in n and m, it follows that the remaining sums in (A.11) give identical expressions. We thus have

$$\Delta^k(u_n v_n) = k! \sum_{n=0}^{k} \sum_{m=0}^{n} \frac{\Delta^n u_n \Delta^m v_n}{(k-n)!(k-m)!(n+m-k)!}$$

$$+ k! \sum_{m=0}^{k} \sum_{n=0}^{m} \frac{\Delta^n u_n \Delta^m v_n}{(k-n)!(k-m)!(n+m-k)!}$$

$$- k! \sum_{n=0}^{k} \frac{\Delta^n u_n \Delta^n v_n}{(k-n)!(k-n)!(2n-k)!} \tag{A.23}$$

These three sums cover the entire region $0 \leq m \leq k, \ 0 \leq n \leq k$, so that we may finally write

$$\Delta^k(u_n v_n) = k! \sum_{n=0}^{k} \sum_{m=0}^{k} \frac{\Delta^n u_n \Delta^m v_n}{(k-n)!(k-m)!(n+m-k)!} \tag{A.24}$$

in which it is understood that terms vanish when $k > n + m$. Alternatively, we may write

$$\Delta^k(u_n v_n) = k! \sum_{n=0}^{k} \sum_{m=0}^{k} \frac{\Delta^{k-n} u_n \Delta^{k-m} v_n}{n!m!(k-n-m)!} \tag{A.25}$$

in which terms vanish when $k < n + m$.

Appendix B
Notation

Throughout this work we encounter matrices of the form

$$
\begin{vmatrix}
u_1 & \cdots & u_{j-1} & 0 & u_{j+1} & \cdots & u_n \\
u_1^{(1)} & \cdots & u_{j-1}^{(1)} & 0 & u_{j+1}^{(1)} & \cdots & u_n^{(1)} \\
\vdots & & \vdots & \vdots & \vdots & & \vdots \\
u_1^{(n-2)} & \cdots & u_{j-1}^{(n-2)} & 0 & u_{j+1}^{(n-2)} & \cdots & u_n^{(n-2)} \\
u_1^{(n-1)} & \cdots & u_{j-1}^{(n-1)} & g_n & u_{j+1}^{(n-1)} & \cdots & u_n^{(n-1)}
\end{vmatrix}
\tag{B.1}
$$

in which the jth column is distinguished, either by having elements which differ from those in the other columns or by being omitted. As written in the above matrix, the meaning is clear for $j = 2, 3, \ldots n - 1$, but not for $j = 1$ or $j = n$. To clarify the intent, which is that the jth column is replaced by something else (or omitted), we give the above matrix for $j = 1$:

$$
\begin{vmatrix}
0 & u_2 & \cdots & u_n \\
0 & u_2^{(1)} & \cdots & u_n^{(1)} \\
\vdots & \vdots & & \vdots \\
0 & u_2^{(n-2)} & \cdots & u_n^{(n-2)} \\
g_n & u_2^{(n-1)} & \cdots & u_n^{(n-1)}
\end{vmatrix}
\tag{B.2}
$$

and for $j = n$:

$$
\begin{vmatrix}
u_1 & \cdots & u_{n-1} & 0 \\
u_1^{(1)} & \cdots & u_{n-1}^{(1)} & 0 \\
\vdots & & \vdots & \vdots \\
u_1^{(n-2)} & \cdots & u_{n-1}^{(n-2)} & 0 \\
u_1^{(n-1)} & \cdots & u_{n-1}^{(n-1)} & g_n
\end{vmatrix}
\tag{B.3}
$$

© Springer International Publishing Switzerland 2016
L.C. Maximon, *Differential and Difference Equations*,
DOI 10.1007/978-3-319-29736-1

For an example in which the jth column is to be omitted, we have the $(n-1) \times (n-1)$ determinant

$$\begin{vmatrix} u_1 & \cdots & u_{j-1} & u_{j+1} & \cdots & u_n \\ u_1^{(1)} & \cdots & u_{j-1}^{(1)} & u_{j+1}^{(1)} & \cdots & u_n^{(1)} \\ \vdots & & \vdots & \vdots & & \vdots \\ u_1^{(n-2)} & \cdots & u_{j-1}^{(n-2)} & u_{j+1}^{(n-2)} & \cdots & u_n^{(n-2)} \end{vmatrix} \tag{B.4}$$

For $j = 1$ this is

$$\begin{vmatrix} u_2 & \cdots & u_n \\ u_2^{(1)} & \cdots & u_n^{(1)} \\ \vdots & \vdots & \vdots \\ u_2^{(n-2)} & \cdots & u_n^{(n-2)} \end{vmatrix} \tag{B.5}$$

and for $j = n$ we have

$$\begin{vmatrix} u_1 & \cdots & u_{n-1} \\ u_1^{(1)} & \cdots & u_{n-1}^{(1)} \\ \vdots & \vdots & \vdots \\ u_1^{(n-2)} & \cdots & u_{n-1}^{(n-2)} \end{vmatrix} \tag{B.6}$$

Appendix C
Wronskian Determinant

The Wronskian determinant for the nth order homogeneous linear differential equation given in (2.1) is defined by the determinant

$$W(x) = \begin{vmatrix} u_1(x) & u_2(x) & \cdots & u_n(x) \\ u_1^{(1)}(x) & u_2^{(1)}(x) & \cdots & u_n^{(1)}(x) \\ \vdots & \vdots & \ddots & \vdots \\ u_1^{(n-1)}(x) & u_2^{(n-1)}(x) & \cdots & u_n^{(n-1)}(x) \end{vmatrix} \tag{C.1}$$

where $u_k(x)$, $k = 1, 2, \ldots n$, are the n linearly independent solutions of (2.1).

The Wronskian determinant obeys the simple first order equation

$$W'(x) = -\frac{a_{n-1}(x)}{a_n(x)} W(x) \tag{C.2}$$

from which, on integrating, we have

$$W(x) = W(x_0) \exp\left(-\int_{x_0}^{x} \frac{a_{n-1}(x')}{a_n(x')} dx'\right), \tag{C.3}$$

known as Abel's theorem.

A variety of derivations of the first order equation for the Wronskian, Eq. (C.2), as well as its integral form, Eq. (C.3), known as Abel's identity, can be found in the literature.[1] We have chosen one that provides as well a derivation of the first derivative of a determinant of any order.[2] It starts with Leibnitz's formula for the expansion of an $n \times n$ determinant, expressed as the sum of $n!$ products of its elements. For the determinant

---

[1] See, e.g., Hartman [20].

[2] A derivation of the expression for the nth derivative of a $j \times j$ determinant has been given by Christiano and Hall, [9].

© Springer International Publishing Switzerland 2016
L.C. Maximon, *Differential and Difference Equations*,
DOI 10.1007/978-3-319-29736-1

$$A(x) = \begin{vmatrix} a_{11}(x) & a_{12}(x) & \ldots & a_{1n}(x) \\ a_{21}(x) & a_{22}(x) & \ldots & a_{2n}(x) \\ \vdots & \vdots & \ddots & \vdots \\ a_{n1}(x) & a_{n2}(x) & \ldots & a_{nn}(x) \end{vmatrix} \qquad \text{(C.4)}$$

the Leibnitz formula is

$$A = \sum_{\sigma \in S_n} \text{sgn}(\sigma) \prod_{i=1}^{n} a_{i\sigma(i)} \qquad \text{(C.5)}$$

Here S_n is the set of all permutations of the integers $\{1, 2, \ldots n\}$, the sum $\sum_{\sigma \in S_n}$ is over all permutations σ, and $\text{sgn}(\sigma)$ is $+1$ for even permutations σ, -1 for odd permutations. In $a_{i\sigma(i)}$, the subscript $\sigma(i)$ is the element in position i in the permutation σ.

Taking the first derivative of A, we have

$$\frac{d}{dx} A = \sum_{\sigma \in S_n} \text{sgn}(\sigma) \sum_{k=1}^{n} a'_{k\sigma(k)} \prod_{\substack{i=1 \\ i \neq k}}^{n} a_{i\sigma(i)}$$

$$= \sum_{k=1}^{n} \sum_{\sigma \in S_n} \text{sgn}(\sigma) \prod_{\substack{i=1 \\ i \neq k}}^{n} a'_{k\sigma(k)} a_{i\sigma(i)} \qquad \text{(C.6)}$$

The derivative $\frac{d}{dx} A$ is thus the sum of n determinants, each obtained by replacing one row in A with the derivative of the elements of that row, leaving all other rows unchanged:

$$\frac{d}{dx} A(x) = \sum_{k=1}^{n} \begin{vmatrix} a_{11}(x) & a_{12}(x) & \ldots & a_{1n}(x) \\ \vdots & \vdots & \ddots & \vdots \\ a'_{k1}(x) & a'_{k2}(x) & \ldots & a'_{kn}(x) \\ \vdots & \vdots & \ddots & \vdots \\ a_{n1}(x) & a_{n2}(x) & \ldots & a_{nn}(x) \end{vmatrix} \qquad \text{(C.7)}$$

Applying Eq. (C.7) for the derivative of a determinant to the Wronskian, Eq. (C.1), we see that the terms in (C.7) for $k = 1, 2, \ldots, n - 1$ vanish since in each of these terms the rows k and $k + 1$ are identical.[3] We are thus left with the term for which $k = n$, giving

---

[3] We use here the property of determinants that if two or more rows (or columns) are identical, then the value of the determinant is zero.

$$\mathcal{W}'(x) = \begin{vmatrix} u_1(x) & u_2(x) & \dots & u_n(x) \\ u_1^{(1)}(x) & u_2^{(1)}(x) & \dots & u_n^{(1)}(x) \\ \vdots & \vdots & \ddots & \vdots \\ u_1^{(n-2)}(x) & u_2^{(n-2)}(x) & \dots & u_n^{(n-2)}(x) \\ u_1^{(n)}(x) & u_2^{(n)}(x) & \dots & u_n^{(n)}(x) \end{vmatrix} \qquad (C.8)$$

From the differential equation, Eq. (2.1), obeyed by each of the functions $u_i(x)$, $\sum_{i=0}^{n} a_i(x)u_k^{(i)}(x) = 0$, $k = 1, 2, \dots, n$, we can express each of the terms in the last row of (C.8) by

$$u_k^{(n)}(x) = -\frac{1}{a_n(x)} \sum_{i=0}^{n-1} a_i(x)u_k^{(i)}(x). \qquad (C.9)$$

Next, multiplying each of the rows corresponding to $i = 0, 1, \dots, n-2$ in (C.9) by $a_i(x)$ and adding them to the last row in (C.8) then cancels all but the term in (C.9) for which $i = n - 1$. The terms in the last row are then $-(a_i(x)/a_n(x))u_k^{(n-1)}(x)$, thus giving the first order differential equation for the Wronskian, Eq. (C.2).[4]

---

[4] Here we have used two other properties of determinants: (1), if one adds to any one row (or column) a linear combination of all other rows (or columns), then the value of a determinant is unchanged, and (2), multiplication of each element in any row (or column) by the same constant multiplies the determinant by that constant.

Appendix D
Casoratian Determinant

The Casoratian determinant for the Nth order homogeneous linear difference equation given in (2.4), analogous to the Wronskian for differential equations, is defined by the determinant

$$
\mathcal{C}(n) = \begin{vmatrix}
u_1(n) & u_2(n) & \cdots & u_N(n) \\
u_1(n+1) & u_2(n+1) & \cdots & u_N(n+1) \\
\vdots & \vdots & \ddots & \vdots \\
u_1(n+N-1) & u_2(n+N-1) & \cdots & u_N(n+N-1)
\end{vmatrix}
\tag{D.1}
$$

where $u_k(n)$, $k = 1, 2, \ldots N$, are the N linearly independent solutions of (2.4).

The Casoratian obeys the simple first order equation

$$
\mathcal{C}(n+1) = (-1)^N \frac{p_0(n)}{p_N(n)} \mathcal{C}(n)
\tag{D.2}
$$

and by iteration we obtain Abel's theorem

$$
\mathcal{C}(n) = (-1)^{N(n-n_0)} \mathcal{C}(n_0) \prod_{j=n_0}^{n-1} \frac{p_0(j)}{p_N(j)}
\tag{D.3}
$$

From (D.1) we have

$$
\mathcal{C}(n+1) = \begin{vmatrix}
u_1(n+1) & u_2(n+1) & \cdots & u_N(n+1) \\
u_1(n+2) & u_2(n+2) & \cdots & u_N(n+2) \\
\vdots & \vdots & \ddots & \vdots \\
u_1(n+N-1) & u_2(n+N-1) & \cdots & u_N(n+N-1) \\
u_1(n+N) & u_2(n+N) & \cdots & u_N(n+N)
\end{vmatrix}
\tag{D.4}
$$

© Springer International Publishing Switzerland 2016
L.C. Maximon, *Differential and Difference Equations*,
DOI 10.1007/978-3-319-29736-1

From the differential equation, Eq. (2.4), obeyed by each of the functions $u_k(n)$, $\sum_{i=0}^{N} p_i(n)u_k(n+i) = 0$, $k = 1, 2, \ldots, N$, we can express each of the terms in the last row of (D.4) by

$$u_k(n+N) = -\frac{1}{p_N(n)} \sum_{i=0}^{N-1} p_i(n)u_k(n+i). \qquad (D.5)$$

Next, multiplying each of the rows corresponding to $i = 1, 2, \ldots, N-1$ in (D.5) by $p_i(n)$ and adding them to the last row in (D.4) then cancels in (D.5) all but the term for which $i = 0$. The terms in the last row are then $-(p_0(n)/p_N(n))u_k(n)$. We now successively interchange the last row with the row above it, moving it finally to the position of first row. This involves $N-1$ interchanges, each of which introduces a factor of -1.[5] The determinant (D.4) is then $(-1)^N p_0(n)/p_N(n)$ times the determinant (D.1), thus giving the first order difference equation for the Casoratian, Eq. (D.2).

---

[5] In addition to the properties of determinants cited in the footnotes in the appendix on the Wronskian, we have used the property that if two adjacent rows (or columns) are interchanged, the determinant is multiplied by -1.

Appendix E
Cramer's Rule

Cramer's rule is an explicit formula for the solution of a system of n linear equations for n unknowns:

$$
\begin{aligned}
a_{11}x_1 + a_{12}x_2 + a_{13}x_3 + \cdots + a_{1n}x_n &= b_1 \\
a_{21}x_1 + a_{22}x_2 + a_{23}x_3 + \cdots + a_{2n}x_n &= b_2 \\
&\vdots \\
a_{n1}x_1 + a_{n2}x_2 + a_{n3}x_3 + \cdots + a_{nn}x_n &= b_n
\end{aligned}
\tag{E.1}
$$

Expressed as a matrix equation, we have

$$
\mathbf{Ax} = \mathbf{b} \tag{E.2}
$$

where

$$
\mathbf{A} = \begin{bmatrix}
a_{11} & a_{12} & \cdots & a_{1n} \\
a_{21} & a_{22} & \cdots & a_{2n} \\
\vdots & \vdots & \ddots & \vdots \\
a_{n1} & a_{n2} & \cdots & a_{nn}
\end{bmatrix}
\tag{E.3}
$$

$$
\mathbf{x} = (x_1, x_2, \ldots, x_n)^{\mathrm{T}} \tag{E.4}
$$

and

$$
\mathbf{b} = (b_1, b_2, \ldots, b_n)^{\mathrm{T}} \tag{E.5}
$$

Cramer's rule expresses the solution, $\mathbf{x} = (x_1, x_2, \ldots, x_n)^{\mathrm{T}}$, in terms of the determinants of the square coefficient matrix $\mathbf{A}$ and of the matrices $\mathbf{A}_i$ formed by replacing the ith column of $\mathbf{A}$ by the column vector $\mathbf{b}$; it is valid and gives a unique solution provided $\det(\mathbf{A}) \neq 0$:

© Springer International Publishing Switzerland 2016
L.C. Maximon, *Differential and Difference Equations*,
DOI 10.1007/978-3-319-29736-1

$$x_i = \frac{\det(\mathbf{A}_i)}{\det(\mathbf{A})} \qquad i = 1, 2, \ldots, n \qquad (E.6)$$

A simple proof of Cramer's rule follows from two properties of determinants: (1) that multiplication of each element in any row (or column) by the same constant multiplies the determinant by that constant, and (2) that adding a constant times any row (or column) to a given row (or column) leaves the value of the determinant unchanged. We illustrate this here for the case of a 3 by 3 matrix and derive the expression for the first unknown, x_1: Here $|\mathbf{A}|$ is the determinant of a 3 by 3 matrix:

$$|\mathbf{A}| = \begin{vmatrix} a_{11} & a_{12} & a_{13} \\ a_{21} & a_{22} & a_{23} \\ a_{31} & a_{32} & a_{33} \end{vmatrix} \qquad (E.7)$$

From the first property we have

$$x_1 |\mathbf{A}| = \begin{vmatrix} a_{11}x_1 & a_{12} & a_{13} \\ a_{21}x_1 & a_{22} & a_{23} \\ a_{31}x_1 & a_{32} & a_{33} \end{vmatrix} \qquad (E.8)$$

and from the second property

$$x_1 |\mathbf{A}| = \begin{vmatrix} a_{11}x_1 + a_{12}x_2 + a_{13}x_3 & a_{12} & a_{13} \\ a_{21}x_1 + a_{22}x_2 + a_{23}x_3 & a_{22} & a_{23} \\ a_{31}x_1 + a_{32}x_2 + a_{33}x_3 & a_{32} & a_{33} \end{vmatrix} = \begin{vmatrix} b_1 & a_{12} & a_{13} \\ b_2 & a_{22} & a_{23} \\ b_3 & a_{32} & a_{33} \end{vmatrix} = |\mathbf{A}_1| \qquad (E.9)$$

Appendix F
Green's Function and the Superposition Principle

In Chap. 6 on Green's function we considered a differential operator L and boundary conditions B_k, operating on a function $y(x)$. Both the differential operator and the boundary conditions are linear operators, from which we have the superposition principle, which simplifies greatly the solution of the differential equation, particularly in the case of general boundary conditions. Although this principle applies equally to the case of an nth order differential equation, we illustrate it here for a second order differential equation.

We consider a function $y_1(x)$ satisfying the inhomogeneous equation $Ly_1(x) = f_1(x)$ and boundary conditions $B_1 y_1 = \gamma_{11}$ and $B_2 y_1 = \gamma_{12}$, and a second function $y_2(x)$ satisfying the inhomogeneous equation $Ly_2(x) = f_2(x)$ and boundary conditions $B_1 y_2 = \gamma_{21}$ and $B_2 y_2 = \gamma_{22}$, with the same differential operator L. For the second order equation in which the function $y(x)$ is considered over the interval $a \le x \le b$ the general boundary conditions B_1 and B_2 are of the form

$$
\begin{aligned}
B_1 y &= \alpha_{11} y(a) + \alpha_{12} y'(a) + \beta_{11} y(b) + \beta_{12} y'(b) = \gamma_1 \\
B_2 y &= \alpha_{21} y(a) + \alpha_{22} y'(a) + \beta_{21} y(b) + \beta_{22} y'(b) = \gamma_2
\end{aligned}
\tag{F.1}
$$

in which α_{jk} and β_{jk} are given constants. Since the differential operator and the boundary conditions are linear, we can write $L(c_1 y_1 + c_2 y_2) = c_1 f_1 + c_2 f_2$ and $B_1(c_1 y_1 + c_2 y_2) = c_1 \gamma_{11} + c_2 \gamma_{12}$ and $B_2(c_1 y_1 + c_2 y_2) = c_1 \gamma_{21} + c_2 \gamma_{22}$, where c_1 and c_2 are constants. This leads directly to the question of the uniqueness of the solution of the inhomogeneous equation: If we consider two solutions, y_1 and y_2, of the equation $Ly = f$ with identical boundary conditions: $B_1 y_1 = \gamma_1$, $B_2 y_1 = \gamma_2$, and $B_1 y_2 = \gamma_1$, $B_2 y_2 = \gamma_2$, then with $c_1 = -1$ and $c_2 = 1$ we obtain the homogeneous equation $L(y_2 - y_1) = 0$ and homogeneous boundary conditions $B_1(y_2 - y_1) = 0$ and $B_2(y_2 - y_1) = 0$. Written more simply, with $Y = y_2 - y_1$ we have $LY = 0$ and $B_1 Y = 0$, $B_2 Y = 0$. Thus if $LY = 0$ has only the trivial solution $Y \equiv 0$ then $y_1 \equiv y_2$, that is, there is at most one unique solution to the equation $Ly = f$ with the boundary conditions $B_1 y = \gamma_1$ and $B_2 y = \gamma_2$. We derive this solution in Chap. 6 on Green's function. On the other hand, if $Y(x)$ has a non-trivial solution then there is either no

© Springer International Publishing Switzerland 2016
L.C. Maximon, *Differential and Difference Equations*,
DOI 10.1007/978-3-319-29736-1

solution to the inhomogeneous equation $Ly = f$ or there are many solutions. We will look into this situation after considering the condition which determines whether there is only the trivial solution $Y \equiv 0$:

Any solution of the equation $LY = 0$ may be written as a sum of two linearly independent solutions of this equation:

$$Y = c_1 u_1 + c_2 u_2$$

Applying the boundary conditions $B_1 Y = B_2 Y = 0$, we have

$$c_1 B_1 u_1 + c_2 B_1 u_2 = 0$$
$$c_1 B_2 u_1 + c_2 B_2 u_2 = 0$$

Therefore, if the determinant

$$\Delta = \begin{vmatrix} B_1 u_1 & B_1 u_2 \\ B_2 u_1 & B_2 u_2 \end{vmatrix} \neq 0 \tag{F.2}$$

then $c_1 = c_2 = 0$ and there is only the trivial solution $Y \equiv 0$. The condition $\Delta \neq 0$ is illustrated if we choose $u_1(x)$ and $u_2(x)$ to be linearly independent solutions of $LY = 0$ such that $B_1 u_1 = B_2 u_2 = 0$. This can be done if we write u_1 and u_2 in terms of two arbitrary linearly independent solutions v_1 and v_2 of $LY = 0$: Defining u_1 and u_2 in terms of v_1 and v_2 by

$$u_1 = (B_1 v_2)v_1 - (B_1 v_1)v_2$$
$$u_2 = (B_2 v_2)v_1 - (B_2 v_1)v_2 \tag{F.3}$$

we then have $B_1 u_1 = B_2 u_2 = 0$ and $B_1 u_2 = -B_2 u_1$, from which $\Delta = (B_1 u_2)^2 \geq 0$.

As an example which shows when there is one solution ($\Delta \neq 0$) and when there is either no solution or many solutions ($\Delta = 0$), we consider the equation $y''(x) + \lambda^2 y(x) = c$ (where c is a constant), with the boundary conditions $B_1 y = y(0) = 0$ and $B_2 y = y(\pi) = 0$. Referring to (F.1), this implies $\alpha_{11} = 1$, $\alpha_{12} = \beta_{11} = \beta_{12} = 0$ and $\beta_{21} = 1$, $\alpha_{21} = \alpha_{22} = \beta_{22} = 0$. If we choose, as solutions of the homogeneous equation $y''(x) + \lambda^2 y(x) = 0$, the two linearly independent functions

$$u_1(x) = \sin \lambda x$$
$$u_2(x) = \cos \lambda \pi \sin \lambda x - \sin \lambda \pi \cos \lambda x \tag{F.4}$$

we then have $B_1 u_1 = u_1(0) = 0$, $B_2 u_2 = u_2(\pi) = 0$ and $B_2 u_1 = \sin \lambda \pi$, $B_1 u_2 = -\sin \lambda \pi$, from which, from (F.2), $\Delta = \sin^2 \lambda \pi$. Thus if $\lambda \neq 1, 2, 3, \ldots$, then $\Delta > 0$ and there is only the trivial solution to $y''(x) + \lambda^2 y(x) = 0$ with boundary conditions $y(0) = y(\pi) = 0$. The inhomogeneous equation $y''(x) + \lambda^2 y(x) = c$ then has the unique solution

$$y(x) = \frac{c}{\lambda^2} \left[(1 - \cos \lambda x) - \frac{(1 - \cos \lambda \pi)}{\sin \lambda \pi} \sin \lambda x \right] \tag{F.5}$$

On the other hand, if $\lambda = 1, 2, 3, \ldots$ then $\Delta = 0$ and the homogeneous equation $y''(x) + \lambda^2 y(x) = 0$ with homogeneous boundary conditions $y(0) = y(\pi) = 0$ has the non-trivial solution $y(x) = A \sin \lambda x$; (λ is an eigenvalue of this equation and $A \sin \lambda x$ is the corresponding eigenfunction). The inhomogeneous equation then has the general solution

$$y(x) = \frac{c}{\lambda^2} [(1 - \cos \lambda x) + A \sin \lambda x] \tag{F.6}$$

where A is an arbitrary constant. Then if $\lambda = 1, 3, 5, \ldots$, the boundary condition $y(\pi) = 0$ can not be satisfied, and there is no solution to the inhomogeneous equation. However, if $\lambda = 2, 4, 6, \ldots$, the general solution just given to the inhomogeneous equation satisfies the boundary conditions with arbitrary A: we have many solutions.

For the nth order equation, the extension of (F.2) giving the necessary and sufficient condition in order that $Ly = f$ with boundary conditions $B_k y = 0$ $(k = 1, 2, \ldots n)$ have a unique non-trivial solution is $\Delta \neq 0$ where

$$\Delta = \begin{vmatrix} B_1 u_1 & B_1 u_2 & \cdots & B_1 u_n \\ B_2 u_1 & B_2 u_2 & \cdots & B_2 u_n \\ \vdots & \vdots & \vdots & \vdots \\ B_n u_1 & B_n u_2 & \cdots & B_n u_n \end{vmatrix} \tag{F.7}$$

(See [33]).

Appendix G
Inverse Laplace Transforms and Inverse Generating Functions

As discussed in Chap. 7 on generating functions, z-transforms and Laplace transforms, the solution of linear differential and difference equations generally requires the inverse of the derived generating function or Laplace transform. In this appendix we derive the inverse of a few generating functions and Laplace transforms that are of particular use in the solution of second order linear differential and difference equations with linear coefficients.

(1) $G(\omega) = \sum_{n=0}^{\infty} y(n)\omega^n = (a - \omega)^{-\alpha}$, in which we assume that α is not a negative integer or zero.

Writing

$$(a - \omega)^{-\alpha} = a^{-\alpha}\left(1 - \frac{\omega}{a}\right)^{-\alpha} = a^{-\alpha}\sum_{n=0}^{\infty}\frac{(\alpha)_n}{n!}\left(\frac{\omega}{a}\right)^n \tag{G.1}$$

we obtain the inverse of the generating function $G(w) = (a - \omega)^{-\alpha}$, namely

$$y(n) = \mathcal{G}^{-1}G(\omega) = a^{-\alpha}\frac{(\alpha)_n}{a^n n!} \tag{G.2}$$

For the particular case in which α is a positive integer: $\alpha = m + 1$, where $m = 0, 1, 2, \ldots$, we have

$$\frac{1}{(a - \omega)^{m+1}} = \sum_{n=0}^{\infty}\frac{\binom{n+m}{m}}{a^{n+m+1}}\omega^n \tag{G.3}$$

and hence

$$y(n) = \mathcal{G}^{-1}\left\{\frac{1}{(a - \omega)^{m+1}}\right\} = \frac{1}{a^{n+m+1}}\binom{n + m}{m} \tag{G.4}$$

as given previously in Eq. (7.56).

© Springer International Publishing Switzerland 2016
L.C. Maximon, *Differential and Difference Equations*,
DOI 10.1007/978-3-319-29736-1

We consider next the generating function

(2) $G(\omega) = (a - \omega)^{-\alpha}(b - \omega)^{-\beta}$, in which we assume that neither α nor β is a negative integer.

We assume further that neither α nor β is zero, and that $a \neq b$, since any of these choices returns us to the simpler generating function just considered. We now have

$$G(\omega) = (a - \omega)^{-\alpha}(b - \omega)^{-\beta} = a^{-\alpha}b^{-\beta} \sum_{l=0}^{\infty} \frac{(\alpha)_l}{l!} \left(\frac{\omega}{a}\right)^l \sum_{m=0}^{\infty} \frac{(\beta)_m}{m!} \left(\frac{\omega}{b}\right)^m \quad (G.5)$$

We then let $n = l + m$, from which

$$(a - \omega)^{-\alpha}(b - \omega)^{-\beta} = a^{-\alpha}b^{-\beta} \sum_{n=0}^{\infty} \frac{\omega^n}{b^n} \sum_{l=0}^{n} \frac{(\alpha)_l (\beta)_{n-l} \, b^l}{l!(n-l)!a^l} \quad (G.6)$$

or, alternatively,

$$(a - \omega)^{-\alpha}(b - \omega)^{-\beta} = a^{-\alpha}b^{-\beta} \sum_{n=0}^{\infty} \frac{\omega^n}{a^n} \sum_{m=0}^{n} \frac{(\beta)_m (\alpha)_{n-m} \, a^m}{m!(n-m)!b^m} \quad (G.7)$$

Here, in (G.6) and (G.7), we write

$$\frac{1}{(n-l)!} = \frac{(-1)^l (-n)_l}{n!}$$

$$(\beta)_{n-l} = (-1)^l \frac{(\beta)_n}{(1 - \beta - n)_l} \quad (G.8)$$

and

$$\frac{1}{(n-m)!} = \frac{(-1)^m (-n)_m}{n!}$$

$$(\alpha)_{n-m} = (-1)^m \frac{(\alpha)_n}{(1 - \alpha - n)_m} \quad (G.9)$$

respectively.

From (G.6) we then have

$$(a - \omega)^{-\alpha}(b - \omega)^{-\beta} = a^{-\alpha}b^{-\beta} \sum_{n=0}^{\infty} \frac{(\beta)_n \omega^n}{b^n n!} \sum_{l=0}^{n} \frac{(-n)_l (\alpha)_l}{l!(1 - \beta - n)_l} \left(\frac{b}{a}\right)^l$$

$$= a^{-\alpha}b^{-\beta} \sum_{n=0}^{\infty} \frac{(\beta)_n}{b^n n!} \, _2F_1\left(-n, \alpha; 1 - \beta - n; \frac{b}{a}\right) \omega^n \quad (G.10)$$

and from (G.7),

$$(a - w)^{-\alpha} (b - w)^{-\beta} = a^{-\alpha} b^{-\beta} \sum_{n=0}^{\infty} \frac{(\alpha)_n w^n}{a^n n!} \sum_{m=0}^{n} \frac{(-n)_m (\beta)_m}{m! (1 - \alpha - n)_m} \left(\frac{a}{b} \right)^m$$

$$= a^{-\alpha} b^{-\beta} \sum_{n=0}^{\infty} \frac{(\alpha)_n}{a^n n!} \, {}_2F_1 \left(-n, \beta; 1 - \alpha - n; \tfrac{a}{b} \right) w^n \qquad \text{(G.11)}$$

The corresponding inverses of the generating function are

$$y(n) = \mathcal{G}^{-1} G(w) = a^{-\alpha} b^{-\beta} \frac{(\beta)_n}{b^n n!} \, {}_2F_1 \left(-n, \alpha; 1 - \beta - n; \tfrac{b}{a} \right) \qquad \text{(G.12)}$$

from (G.10), and

$$y(n) = \mathcal{G}^{-1} G(w) = a^{-\alpha} b^{-\beta} \frac{(\alpha)_n}{a^n n!} \, {}_2F_1 \left(-n, \beta; 1 - \alpha - n; \tfrac{a}{b} \right) \qquad \text{(G.13)}$$

from (G.11).

Here (G.12) is valid if either $\beta \neq$ integer or $\beta = 2, 3, \ldots$, (i.e., a positive integer >1), and α is arbitrary; (G.13) is valid if either $\alpha \neq$ integer or $\alpha = 2, 3, \ldots$, (i.e., a positive integer >1), and β is arbitrary. Thus for the following choices for α and β, we may choose either (G.12) or (G.13), or both, as solutions.

$$
\begin{array}{ll}
\alpha \neq \text{integer}, \ \beta \neq \text{integer}: & \text{(G.12) or (G.13).} \\
\alpha \neq \text{integer}, \ \beta \ \text{arbitary}: & \text{(G.13).} \\
\beta \neq \text{integer}, \ \alpha \ \text{arbitary}: & \text{(G.12).} \\
\alpha, \ \beta = 2, 3, \ldots \text{(i.e., both positive integers} >1): & \text{(G.12) or (G.13).} \\
\alpha = 1, \ \beta = 2, 3, \ldots \text{(i.e., a positive integer} >1): & \text{(G.12).} \\
\beta = 1, \ \alpha = 2, 3, \ldots \text{(i.e., a positive integer} >1): & \text{(G.13).}
\end{array}
$$
(G.14)

We consider next the Laplace transform
(3) $F(s) = (s - \alpha)^{-\beta}$.
From [13, Sect. 1.1 (5)],

$$\int_0^{\infty} e^{-sx} x^{\beta - 1} \, dx = \Gamma(\beta) s^{-\beta} \qquad \Re \, \beta > 0, \qquad \text{(G.15)}$$

so that

$$\frac{1}{\Gamma(\beta)} \int_0^{\infty} e^{-sx} e^{\alpha x} x^{\beta - 1} \, dx = (s - \alpha)^{-\beta} \qquad \text{(G.16)}$$

We thus have the inverse transform of $F(s) = (s - \alpha)^{-\beta}$, namely

$$\mathcal{L}^{-1} F(s) = \frac{1}{2\pi i} \int_{\gamma - i\infty}^{\gamma + i\infty} \frac{e^{sx}}{(s - \alpha)^\beta} \, ds = \frac{x^{\beta - 1}}{\Gamma(\beta)} e^{\alpha x} \qquad\qquad \text{(G.17)}$$

Next we consider the Laplace transform
(4) $F(s) = (s - \alpha_1)^{-\beta_1}(s - \alpha_2)^{-\beta_2}$.
The inverse Laplace transform is

$$\begin{aligned}
\mathcal{L}^{-1} F(s) &= \frac{1}{2\pi i} \int_{\gamma - i\infty}^{\gamma + i\infty} \frac{e^{sx}}{(s - \alpha_1)^{\beta_1}(s - \alpha_2)^{\beta_2}} \, ds \\
&= \frac{1}{2\pi i} e^{\alpha_1 x} \int_{\gamma - i\infty}^{\gamma + i\infty} e^{sx} s^{-\beta_1} (s - (\alpha_2 - \alpha_1))^{-\beta_2} \, ds \\
&= \frac{1}{2\pi i} (\alpha_2 - \alpha_1)^{-\beta_1 - \beta_2 + 1} e^{\alpha_1 x} \int_{\gamma - i\infty}^{\gamma + i\infty} e^{(\alpha_2 - \alpha_1)x\sigma} \sigma^{-\beta_1} (\sigma - 1)^{-\beta_2} \, d\sigma \\
&= \frac{1}{2\pi i} (\alpha_2 - \alpha_1)^{-\beta_1 - \beta_2 + 1} e^{\alpha_1 x} \int_{\gamma - i\infty}^{\gamma + i\infty} e^{(\alpha_2 - \alpha_1)x\sigma} \sigma^{-\beta_1 - \beta_2} (1 - \sigma^{-1})^{-\beta_2} \, d\sigma
\end{aligned}$$

$$\text{(G.18)}$$

From [13, Sect. 6.10(6)],

$$\int_0^\infty e^{-sx} x^{c-1} \, {}_1F_1(a; c; x) \, dx = \Gamma(c) s^{-c} (1 - s^{-1})^{-a} \qquad \Re c > 0, \ \Re s > 1,$$

$$\text{(G.19)}$$

from which we have the inverse transform

$$\frac{1}{2\pi i} \Gamma(c) \int_{\gamma - i\infty}^{\gamma + i\infty} e^{sx} s^{-c} (1 - s^{-1})^{-a} \, ds = x^{c-1} \, {}_1F_1(a; c; x) \qquad\qquad \text{(G.20)}$$

Thus,

$$\begin{aligned}
\mathcal{L}^{-1} F(s) &= \frac{1}{2\pi i} \int_{\gamma - i\infty}^{\gamma + i\infty} \frac{e^{sx}}{(s - \alpha_1)^{\beta_1}(s - \alpha_2)^{\beta_2}} \, ds \\
&= \frac{x^{\beta_1 + \beta_2 - 1}}{\Gamma(\beta_1 + \beta_2)} e^{\alpha_1 x} \, {}_1F_1(\beta_2; \beta_1 + \beta_2; (\alpha_2 - \alpha_1)x) \\
&= \frac{x^{\beta_1 + \beta_2 - 1}}{\Gamma(\beta_1 + \beta_2)} e^{\alpha_2 x} \, {}_1F_1(\beta_1; \beta_1 + \beta_2; (\alpha_1 - \alpha_2)x)
\end{aligned}$$

$$\text{(G.21)}$$

Appendix H
Hypergeometric Function

In this appendix we give a few of the transformations of the hypergeometric function $_2F_1(a, b; c; z)$ which have been useful in the analysis presented in this work.

$$_2F_1(a, b; c; z) = {}_2F_1(b, a; c; z) \tag{H.1}$$

$$\begin{aligned}
_2F_1(a, b; c; z) &= (1 - z)^{-a}{}_2F_1\left(a, c - b; c; \tfrac{z}{z-1}\right) \\
&= (1 - z)^{-b}{}_2F_1\left(c - a, b; c; \tfrac{z}{z-1}\right)
\end{aligned} \tag{H.2}$$

$$_2F_1(a, b; c; z) = (1 - z)^{c-a-b}{}_2F_1(c - a, c - b; c; z) \tag{H.3}$$

For (H.1) see [13, Sect. 2.1.2, p. 57]; for (H.2) and (H.3) see [13, Sect. 2.1.4, (22) and (23), p. 64].

$$\begin{aligned}
_2F_1(a, b; c; z) &= e^{i\pi a}\frac{\Gamma(c)\Gamma(b - c + 1)}{\Gamma(a + b - c + 1)\Gamma(c - a)}z^{-a}{}_2F_1\left(a, a - c + 1; a + b - c + 1; 1 - \tfrac{1}{z}\right) \\
&\quad + \frac{\Gamma(c)\Gamma(b - c + 1)}{\Gamma(a)\Gamma(b - a + 1)}z^{a-c}(1 - z)^{c-a-b}{}_2F_1\left(1 - a, c - a; b - a + 1; \tfrac{1}{z}\right) \\
&= e^{i\pi b}\frac{\Gamma(c)\Gamma(a - c + 1)}{\Gamma(a + b - c + 1)\Gamma(c - b)}z^{-b}{}_2F_1\left(b, b - c + 1; a + b - c + 1; 1 - \tfrac{1}{z}\right) \\
&\quad + \frac{\Gamma(c)\Gamma(a - c + 1)}{\Gamma(b)\Gamma(a - b + 1)}z^{b-c}(1 - z)^{c-a-b}{}_2F_1\left(1 - b, c - b; a - b + 1; \tfrac{1}{z}\right)
\end{aligned} \tag{H.4}$$

For (H.4) see [40, Sect. 4, (26), p. 447] or [13, Sect. 2.9, (26), p. 106].

© Springer International Publishing Switzerland 2016
Ł.C. Maximon, *Differential and Difference Equations*,
DOI 10.1007/978-3-319-29736-1

Appendix I
Confluent Hypergeometric Functions

In this appendix we list the confluent hypergeometric functions which result from different choices of integration path for the integrals given earlier in Eqs. (7.200)–(7.203). Here, from (7.192)–(7.194), (7.199), and (7.206),

$$\gamma_i = \frac{c_i}{d_i}, \qquad \beta_0 = \gamma_0, \qquad \beta_1 = \gamma_0 - \gamma_2, \qquad z = \gamma_0 - 2\gamma_1 + \gamma_2$$

From (7.200), we have the integral

$$\int s^{\beta_1-\beta_0-x-1}(1-s)^{\beta_0+x-1}e^{zs}\,ds = \int s^{-\gamma_2-x-1}(1-s)^{\gamma_0+x-1}e^{zs}\,ds$$

leading to the confluent hypergeometric function $_1F_1$:

$$
\begin{aligned}
\int_0^1 s^{-\gamma_2-x-1}(1-s)^{\gamma_0+x-1}e^{zs}\,ds &= \frac{\Gamma(-\gamma_2-x)\Gamma(\gamma_0+x)}{\Gamma(\beta_1)}{}_1F_1(-\gamma_2-x;\beta_1;z) \\
&= \frac{\Gamma(-\gamma_2-x)\Gamma(\gamma_0+x)}{\Gamma(\beta_1)}e^z{}_1F_1(\gamma_0+x;\beta_1;-z) \qquad \text{(I.1)}
\end{aligned}
$$

$$
\begin{aligned}
\int_0^{(1+)} s^{-\gamma_2-x-1}(1-s)^{\beta_0+x-1}e^{zs}\,ds &= 2\pi i\,e^{(\gamma_0+x-1)\pi i}\frac{\Gamma(-\gamma_2-x)}{\Gamma(1-\gamma_0-x)\Gamma(\beta_1)}{}_1F_1(-\gamma_2-x;\beta_1;z) \\
&= 2\pi i\,e^{(\gamma_0+x-1)\pi i}\frac{\Gamma(-\gamma_2-x)}{\Gamma(1-\gamma_0-x)\Gamma(\beta_1)}e^z{}_1F_1(\gamma_0+x;\beta_1;-z) \\
&\hspace{9cm} \text{(I.2)}
\end{aligned}
$$

$$
\begin{aligned}
\int_1^{(0+)} s^{-\gamma_2-x-1}(1-s)^{\gamma_0+x-1}e^{zs}\,ds &= 2\pi i\,e^{(-\gamma_2-x)\pi i}\frac{\Gamma(\gamma_0+x)}{\Gamma(\gamma_2+x+1)\Gamma(\beta_1)}{}_1F_1(-\gamma_2-x;\beta_1;z) \\
&= 2\pi i\,e^{(\beta_1-\beta_0-x)\pi i}\frac{\Gamma(\gamma_0+x)}{\Gamma(\gamma_2+x+1)\Gamma(\beta_1)}e^z{}_1F_1(\gamma_0+x;\beta_1;-z) \\
&\hspace{9cm} \text{(I.3)}
\end{aligned}
$$

© Springer International Publishing Switzerland 2016
L.C. Maximon, *Differential and Difference Equations*,
DOI 10.1007/978-3-319-29736-1

$$\int_{\alpha}^{(0+,1+,0-,1-)} s^{-(\gamma_2+x+1)}(1-s)^{(\gamma_0+x-1)}e^{zs}\,ds = 4\pi^2 \frac{e^{\beta_1\pi i}\,_1F_1(-\gamma_2-x;\beta_1;z)}{\Gamma(\gamma_2+x+1)\Gamma(1-\gamma_0-x)\Gamma(\beta_1)}$$

$$= 4\pi^2 \frac{e^{\beta_1\pi i}e^z\,_1F_1(\gamma_0+x;\beta_1;-z)}{\Gamma(\gamma_2+x+1)\Gamma(1-\gamma_0-x)\Gamma(\beta_1)} \tag{I.4}$$

From (7.201), we have

$$\int_{-\infty}^{(0+,1+)} s^{-(2-\beta_1)}(1-\tfrac{1}{s})^{\beta_0+x-1}e^{zs}\,ds = 2\pi i\,z^{1-\beta_1}\frac{1}{\Gamma(2-\beta_1)}\,_1F_1(1-\gamma_0-x;2-\beta_1;z)$$

$$= 2\pi i\,z^{1-\beta_1}\frac{1}{\Gamma(2-\beta_1)}e^z\,_1F_1(\gamma_2+x+1;2-\beta_1;-z) \tag{I.5}$$

Here, in each of the last five equations, the second confluent hypergeometric function is obtained using the Kummer transformation ([36, Sect. 13.2(vii), Eq. 13.2.39]):

$$_1F_1(a;c;z) = e^z\,_1F_1(c-a;c;-z) \tag{I.6}$$

From (7.202), we have, for the integral $\int s^{\beta_1-\beta_0-x-1}(1+s)^{\beta_0+x-1}e^{-zs}\,ds$ leading to confluent hypergeometric functions of the form $U(a;c;z)$,

$$\int_0^\infty s^{\beta_1-\beta_0-x-1}(1+s)^{\beta_0+x-1}e^{-zs}\,ds = \Gamma(-\gamma_2-x)U(-\gamma_2-x;\beta_1;z)$$

$$= \Gamma(-\gamma_2-x)z^{1-\beta_1}U(1-\gamma_0-x;2-\beta_1;z) \tag{I.7}$$

$$\int_\infty^{(0+)} s^{\beta_1-\beta_0-x-1}(1+s)^{\beta_0+x-1}e^{-zs}\,ds = 2\pi i\frac{e^{(-\gamma_2-x)\pi i}}{\Gamma(\gamma_2+x+1)}U(-\gamma_2-x;\beta_1;z)$$

$$= 2\pi i\frac{e^{(-\gamma_2-x)\pi i}}{\Gamma(\gamma_2+x+1)}z^{1-\beta_1}U(1-\gamma_0-x;2-\beta_1;z) \tag{I.8}$$

and from (7.203), we have, for the integral $\int s^{\beta_0+x-1}(1+s)^{\beta_1-\beta_0-x-1}e^{zs}\,ds$,

$$\int_0^\infty s^{\beta_0+x-1}(1+s)^{\beta_1-\beta_0-x-1}e^{zs}\,ds = \Gamma(\gamma_0+x)U(\gamma_0+x;\beta_1;-z)$$

$$= \Gamma(\gamma_0+x)(-z)^{1-\beta_1}U(\gamma_2+x+1;2-\beta_1;-z) \tag{I.9}$$

$$\int_\infty^{(0+)} s^{\beta_0+x-1}(1+s)^{\beta_1-\beta_0-x-1}e^{zs}\,ds = 2\pi i\frac{e^{(\gamma_0+x)\pi i}}{\Gamma(1-\gamma_0-x)}U(\gamma_0+x;\beta_1;-z)$$

$$= 2\pi i\frac{e^{(\gamma_0+x)\pi i}}{\Gamma(1-\gamma_0-x)}(-z)^{1-\beta_1}U(\gamma_2+x+1;2-\beta_1;-z) \tag{I.10}$$

Here, in each of the last four equations, the second confluent hypergeometric function is obtained using the Kummer transformation ([36, Sect. 13.2(vii), Eq. 13.2.40]):

$$U(a; c; z) = z^{1-c} U(a - c + 1; 2 - c; z) \tag{I.11}$$

In order to obtain a solution to the difference equation that satisfies arbitrary initial conditions $y(x_0)$ and $y(x_0 + 1)$, we require two linearly independent solutions of the difference equation, $y_1(x)$ and $y_2(x)$. We therefore wish to chose, from among the solutions listed above, pairs of solutions which are linearly independent. The particular pair of solutions chosen will depend on the value of the parameters $\gamma_2 + x + 1$ and $\gamma_0 + x - 1$. It is important to note, however, that the recursion defined by (7.205) fails if $\gamma_2 + x + 1 = 0$ if we assume to have two independent initial conditions, for example $w(0)$ and $w(1)$. The condition that the two solutions to (7.205) be linearly independent is that their Casoratian, $\mathcal{C}(x)$, be non-zero: $\mathcal{C}(x) = y_1(x) y_2(x + 1) - y_1(x + 1) y_2(x) \neq 0$. Noting that the Wronskian of the various solutions that we are considering is relatively well-known, (see, e.g., [36, Sect. 13.2(vi)]), we determine the Casoratian by expressing it in terms of the Wronskian by the use of raising and lowering operators; these relate the differential properties of the variable z in the solution to the discrete property of the parameter x.

From (I.1)–(I.10) we define the ten functions $F1 - F5t$ of the form $_1F_1(a; c; z)$ and the eight functions $U1 - U4t$ of the form $U(a; c; z)$, from which we obtain linearly independent pairs of solutions of the difference equation (7.153) in which $d_1^2 - 4d_0d_2 = 0$, each pair being valid for given values of the variable z and the parameters $\gamma_0 + x$ and $\gamma_2 + x + 1$, presented in detail in Table 7.1. As noted previously, in defining $F1 - F5t$ and $U1 - U4t$ we have neglected factors independent of x but retain the gamma functions $\Gamma(\beta_1)$ and $\Gamma(2 - \beta_1)$ in the denominators; the functions of the form $_1F_1(a; c; z)/\Gamma(c)$ then remain well-defined when the parameter c is zero or a negative integer. We then have, from (I.1)–(I.10),

$$F1 = \frac{\Gamma(-\gamma_2 - x)\Gamma(\gamma_0 + x)}{\Gamma(\beta_1)} e^{i\pi x} {}_1F_1(-\gamma_2 - x; \beta_1; z) \tag{I.12}$$

$$F1t = \frac{\Gamma(-\gamma_2 - x)\Gamma(\gamma_0 + x)}{\Gamma(\beta_1)} e^{i\pi x} {}_1F_1(\gamma_0 + x; \beta_1; -z) \tag{I.13}$$

$$F2 = \frac{\Gamma(-\gamma_2 - x)}{\Gamma(1 - \gamma_0 - x)\Gamma(\beta_1)} {}_1F_1(-\gamma_2 - x; \beta_1; z) \tag{I.14}$$

$$F2t = \frac{\Gamma(-\gamma_2 - x)}{\Gamma(1 - \gamma_0 - x)\Gamma(\beta_1)} {}_1F_1(\gamma_0 + x; \beta_1; -z) \tag{I.15}$$

$$F3 = \frac{\Gamma(\gamma_0 + x)}{\Gamma(\gamma_2 + x + 1)\Gamma(\beta_1)} {}_1F_1(-\gamma_2 - x; \beta_1; z) \tag{I.16}$$

$$F3t = \frac{\Gamma(\gamma_0 + x)}{\Gamma(\gamma_2 + x + 1)\Gamma(\beta_1)} {}_1F_1(\gamma_0 + x; \beta_1; -z) \tag{I.17}$$

$$F4 = \frac{e^{i\pi x}}{\Gamma(\gamma_2 + x + 1)\Gamma(1 - \gamma_0 - x)} \frac{{}_1F_1(-\gamma_2 - x; \beta_1; z)}{\Gamma(\beta_1)} \tag{I.18}$$

$$F4t = \frac{e^{i\pi x}}{\Gamma(\gamma_2 + x + 1)\Gamma(1 - \gamma_0 - x)} \frac{{}_1F_1(\gamma_0 + x; \beta_1; -z)}{\Gamma(\beta_1)} \tag{I.19}$$

$$F5 = \frac{{}_1F_1(1 - \gamma_0 - x; 2 - \beta_1; z)}{\Gamma(2 - \beta_1)} \tag{I.20}$$

$$F5t = \frac{{}_1F_1(\gamma_2 + x + 1; 2 - \beta_1; -z)}{\Gamma(2 - \beta_1)} \tag{I.21}$$

$$U1 = \Gamma(-\gamma_2 - x)U(-\gamma_2 - x; \beta_1; z) \tag{I.22}$$

$$U1t = \Gamma(-\gamma_2 - x)U(1 - \gamma_0 - x; 2 - \beta_1; z) \tag{I.23}$$

$$U2 = \frac{e^{i\pi x}}{\Gamma(\gamma_2 + x + 1)}U(-\gamma_2 - x; \beta_1; z) \tag{I.24}$$

$$U2t = \frac{e^{i\pi x}}{\Gamma(\gamma_2 + x + 1)}U(1 - \gamma_0 - x; 2 - \beta_1; z) \tag{I.25}$$

$$U3 = \Gamma(\gamma_0 + x)U(\gamma_0 + x; \beta_1; -z) \tag{I.26}$$

$$U3t = \Gamma(\gamma_0 + x)U(\gamma_2 + x + 1; 2 - \beta_1; -z) \tag{I.27}$$

$$U4 = \frac{e^{i\pi x}}{\Gamma(1 - \gamma_0 - x)}U(\gamma_0 + x; \beta_1; -z) \tag{I.28}$$

$$U4t = \frac{e^{i\pi x}}{\Gamma(1 - \gamma_0 - x)}U(\gamma_2 + x + 1; 2 - \beta_1; -z) \tag{I.29}$$

Appendix J
Solutions of the Second Kind

In this appendix we present two linearly independent solutions of the difference equation given in (7.205), namely

$$(\gamma_2 + x + 1)\, w(x + 1) - 2\,(\gamma_1 + x)\, w(x) + (\gamma_0 + x - 1)\, w(x - 1) = 0, \quad \text{(J.1)}$$

for the case in which $\gamma_2 + x + 1$ and $\gamma_0 + x$ are positive integers, from which $\beta_1 \equiv \gamma_0 - \gamma_2$ is an integer. Referring to Table 7.1, we consider first the case in which $\beta_1 \neq 0, -1, -2, \ldots$ (and hence $\beta_1 = 1, 2, 3, \ldots$). Then, for $z \equiv \gamma_0 - 2\gamma_1 + \gamma_2 < 0$, two linearly independent solutions are

$$U3 = \Gamma(\gamma_0 + x)U(\gamma_0 + x; \beta_1; -z) \quad \text{(J.2)}$$

and

$$F3t = \frac{\Gamma(\gamma_0 + x)}{\Gamma(\gamma_2 + x + 1)\Gamma(\beta_1)}\, {}_1F_1(\gamma_0 + x; \beta_1; -z) \quad \text{(J.3)}$$

Now from [36, Sect. 13.2(i), Eq. 13.2.9], with $\beta_1 = n + 1, a = \gamma_0 + x$, and $a - n = \gamma_2 + x + 1$

$$
\begin{aligned}
U3 &= \Gamma(\gamma_0 + x)U(\gamma_0 + x; \beta_1; -z) \\
&= \frac{(-1)^{n+1}\Gamma(\gamma_0 + x)}{n!\,\Gamma(\gamma_2 + x + 1)} \sum_{k=0}^{\infty} \frac{(\gamma_0 + x)_k(-z)^k}{(n+1)_k k!} \left\{ \ln(-z) + \Psi(\gamma_0 + x + k) - \Psi(1 + k) - \Psi(n + 1 + k) \right\} \\
&\quad + \sum_{k=1}^{n} \frac{(k-1)!\,(1 + k - \gamma_0 - x)_{n-k}}{(n-k)!} (-z)^{-k} \\
&= (-1)^{n+1} F3t \ln(-z) \\
&\quad + \frac{(-1)^{n+1}\Gamma(\gamma_0 + x)}{n!\,\Gamma(\gamma_2 + x + 1)} \sum_{k=0}^{\infty} \frac{(\gamma_0 + x)_k(-z)^k}{(n+1)_k k!} \left\{ \Psi(\gamma_0 + x + k) - \Psi(1 + k) - \Psi(n + 1 + k) \right\} \\
&\quad + \sum_{k=1}^{n} \frac{(k-1)!\,(1 + k - \gamma_0 - x)_{n-k}}{(n-k)!} (-z)^{-k} \quad \text{(J.4)}
\end{aligned}
$$

© Springer International Publishing Switzerland 2016
L.C. Maximon, *Differential and Difference Equations*,
DOI 10.1007/978-3-319-29736-1

Here, since both U3 and F3t are solutions of the difference equation for $w(x)$, the remaining terms on the right hand side of the above equation also satisfy this equation, and give a linearly independent solution to that equation.

Furthermore, these terms give a valid solution for both $z < 0$ and $z > 0$. Thus, for $\beta_1 \equiv \gamma_0 - \gamma_2 = 1 + n = 1, 2, 3, \ldots$, two linearly independent solutions, valid for all z, are F3t and

$$\frac{(-1)^{n+1} \Gamma(\gamma_0 + x)}{n! \Gamma(\gamma_2 + x + 1)} \sum_{k=0}^{\infty} \frac{(\gamma_0 + x)_k (-z)^k}{(n+1)_k k!} \left\{ \Psi(\gamma_0 + x + k) - \Psi(1+k) - \Psi(n+1+k) \right\}$$

$$+ \sum_{k=1}^{n} \frac{(k-1)!(1+k-\gamma_0 - x)_{n-k}}{(n-k)!} (-z)^{-k} \tag{J.5}$$

The case in which $\beta_1 \neq 2, 3, 4, \ldots$, (and hence $\beta_1 = 1, 0, -1, -2, \ldots$), may be treated in similar fashion. For $z < 0$, two linearly independent solutions are

$$U3t = \Gamma(\gamma_0 + x) U(\gamma_2 + x + 1; 2 - \beta_1; -z) \tag{J.6}$$

and

$$F5t = \frac{{}_1F_1(\gamma_2 + x + 1; 2 - \beta_1; -z)}{\Gamma(2 - \beta_1)} \tag{J.7}$$

Again from [36, Sect. 13.2(i), Eq. 13.2.9], now with $2 - \beta_1 = n + 1$, $a = \gamma_2 + x + 1$, and $a - n = \gamma_0 + x$,

$$U3t = \Gamma(\gamma_0 + x) U(\gamma_2 + x + 1; 2 - \beta_1; -z)$$

$$= \frac{(-1)^{n+1}}{n!} \sum_{k=0}^{\infty} \frac{(\gamma_2 + x + 1)_k (-z)^k}{(n+1)_k k!} \left\{ \ln(-z) + \Psi(\gamma_2 + x + 1 + k) - \Psi(1+k) - \Psi(n+1+k) \right\}$$

$$+ \frac{\Gamma(\gamma_0 + x)}{\Gamma(\gamma_2 + x + 1)} \sum_{k=1}^{n} \frac{(k-1)!(k - \gamma_2 - x)_{n-k}}{(n-k)!} (-z)^{-k}$$

$$= (-1)^{n+1} F5t \ln(-z)$$

$$+ \frac{(-1)^{n+1}}{n!} \sum_{k=0}^{\infty} \frac{(\gamma_2 + x + 1)_k (-z)^k}{(n+1)_k k!} \left\{ \Psi(\gamma_2 + x + 1 + k) - \Psi(1+k) - \Psi(n+1+k) \right\}$$

$$+ \frac{\Gamma(\gamma_0 + x)}{\Gamma(\gamma_2 + x + 1)} \sum_{k=1}^{n} \frac{(k-1)!(k - \gamma_2 - x)_{n-k}}{(n-k)!} (-z)^{-k} \tag{J.8}$$

Here, since both U3t and F5t are solutions of the difference equation for $w(x)$, the remaining terms on the right hand side of the above equation also satisfy this equation, and give a linearly independent solution to that equation. Furthermore, these terms give a valid solution for both $z < 0$ and $z > 0$. Thus, for $\beta_1 \equiv \gamma_0 - \gamma_2 = 1 - n = 1, 0, -1, -2, \ldots$, two linearly independent solutions, valid for all z, are F5t and

$$\frac{(-1)^{n+1}}{n!} \sum_{k=0}^{\infty} \frac{(\gamma_2 + x + 1)_k (-z)^k}{(n+1)_k k!} \{\Psi(\gamma_2 + x + 1 + k) - \Psi(1+k) - \Psi(n+1+k)\}$$

$$+ \frac{\Gamma(\gamma_0 + x)}{\Gamma(\gamma_2 + x + 1)} \sum_{k=1}^{n} \frac{(k-1)!(k - \gamma_2 - x)_{n-k}}{(n-k)!} (-z)^{-k} \tag{J.9}$$

An alternative approach which derives a second solution in the form of a polynomial in x is given in [37]. For $\beta_1 \neq 0, -1, -2, \ldots$, we write $\beta_1 = n + 1 = 1, 2, \ldots$; $\gamma_2 + x = N = 0, 1, 2, \ldots$ and choose F3 as the first solution to the difference equation, writing

$$F3 = \frac{\Gamma(\gamma_0 + x)}{\Gamma(\gamma_2 + x + 1)\Gamma(\beta_1)} {}_1F_1(-\gamma_2 - x; \beta_1; z)$$

$$= \frac{(N+n)!}{n!N!} {}_1F_1(-N; n+1; z) \tag{J.10}$$

We note that the function given here is the associated Laguerre polynomial (see [36, Sect. 13.6(v), Eq. 13.6.19 and Sect. 18.5(iii), Eq. 18.5.12]):

$$\frac{(N+n)!}{n!N!} {}_1F_1(-N; n+1; z) = L_N^{(n)}(z). \tag{J.11}$$

As shown in [34, Sect. 11, Eq. (4), p. 97], given a polynomial solution $y_n(z)$ to a differential equation of the hypergeometric type,[6] a solution of the second kind may be obtained as an extended Cauchy-integral:

$$Q_n(z) = \frac{1}{\rho(z)} \int_0^{\infty} \frac{y_n(s)\rho(s)}{s - z} ds, \tag{J.12}$$

where, for the confluent hypergeometric function ${}_1F_1(-N; n+1; z)$, satisfying the differential equation $zy''(z) + (n + 1 - z)y'(z) + Ny(z) = 0$ considered here, $\rho(z) = z^n e^{-z}$ is a solution of $(z\rho(z))' = (n + 1 - z)\rho(z)$. This approach is developed in detail in [37], leading to a second linearly independent polynomial solution in closed form which satisfies both the differential equation and the difference equation (7.205) in each of the parameters. From [37, Eq. (2.12)], we then have two linearly independent polynomial solutions of the difference equation (7.205), in which $\gamma_2 + x = N; n + 1 = \beta_1 = \gamma_0 - \gamma_2 \neq 0, -1, -2, \ldots$ and $z = \gamma_0 - 2\gamma_1 + \gamma_2$:

$$\frac{(N+n)!}{n!N!} {}_1F_1(-N; n+1; z) \tag{J.13}$$

and

---

[6]I.e., $\sigma(z)y''(z) + \tau(z)y'(z) + \lambda y(z) = 0$. See [34, Sects. 2, 3, pp. 6–14].

$$\frac{(N+n)!}{n!N!} \sum_{k=0}^{N} \sum_{m=0}^{n+k-1} \frac{(-N)_k(n+k-1-m)!}{(n+1)_k \, k!} x^m . \tag{J.14}$$

For the case in which $\beta_1 \neq 2, 3, \ldots$ we write $2 - \beta_1 = n + 1 = 1, 2, \ldots$; $\gamma_0 + x - 1 = N = 0, 1, 2, \ldots$ and choose F5 as the first solution to the difference equation:

$$F5 = \frac{{}_1F_1(1 - \gamma_0 - x; 2 - \beta_1; z)}{\Gamma(2 - \beta_1)}$$

$$= \frac{{}_1F_1(-N; n+1; z)}{n!} \tag{J.15}$$

The second, linearly independent solution of (7.205), is then

$$\frac{1}{n!} \sum_{k=0}^{N} \sum_{m=0}^{n+k-1} \frac{(-N)_k(n+k-1-m)!}{(n+1)_k \, k!} x^m . \tag{J.16}$$

Our second polynomial solution enables us to define a linearly independent associated Laguerre function of the second kind which satisfies the difference equations for the confluent hypergeometric function in each of its two parameters in terms of appropriately normalized polynomials. (See [37]).

Bibliography

1. Abdel-Messih MA (1959) A Green's function analog for ordinary linear difference equations. Proc Math Phys Soc Egypt 22(1958):43–51
2. Agarwal RP (2000) Difference equations and inequalities: theory, methods and applications, 2nd edn, revised and expanded. Marcel Dekker Inc., New York
3. Arfken G (1985) Mathematical methods for physicists, 3rd edn. Academic Press, New York
4. Atkinson FV (1964) Discrete and continuous boundary problems. Academic Press, New York
5. Batchelder PM (1927) An introduction to linear difference equations. The Harvard University Press, Cambridge
6. Bôcher M (1911) Boundary problems and Green's functions for linear differential and difference equations. Ann Math Ser. 2, 13(1):71–88
7. Bôcher M (1912) Problems boundary, in one dimension (a lecture delivered 27 Aug 1912). In: Proceedings of the fifth international congress of mathematicians, 22–28 Aug 1912, vol 1 (1913). Cambridge, England, University Press, pp 163–195
8. Boyce WE, DiPrima RC (2013) Elementary differential equations and boundary value problems. Wiley, New York
9. Christiano JG, Hall JE (1964) On the nth derivative of a determinant of the jth order. Math Mag 37(4):215–217
10. Coddington EA, Levinson N (1955) Theory of ordinary differential equations. McGraw Hill, New York
11. Dennery P, Krzywicki A (1967) Mathematics for physicists. Harper & Row, New York
12. Elaydi SN (2005) An introduction to difference equations, 3rd edn. Springer, New York
13. Erdélyi A (ed) (1953) Higher transcendental functions, vol 1. McGraw-Hill Book Company, New York
14. Feigenbaum MJ (1978) Quantitative universality for a class of nonlinear transformations. J Stat Phys 19(1):25–52
15. Forsyth AR (1959) Theory of differential equations, Part III, Ordinary linear equations. Dover Publications Inc., New York (Note Chapter VII, in particular Secs. 101–108)
16. Fort T (1948) Finite differences and difference equations in the real domain. Oxford University Press, London
17. Fowler D, Robson E (1998) Square root approximations in Old Babylonian mathematics: YBC 7289 in context. Hist Math 25:366–378, article no. HM982209
18. Golomb M, Shanks M (1965) Elements of ordinary differential equations. McGraw-Hill, New York
19. Hairer E, Nørsett SP, Wanner G (1993) Solving ordinary differential equations I, Nonstiff problems, 2nd revised edn. Springer, Berlin

© Springer International Publishing Switzerland 2016

L.C. Maximon, *Differential and Difference Equations*,
DOI 10.1007/978-3-319-29736-1

20. Hartman P (1964) Ordinary differential equations. Wiley, New York
21. Heath TL (1921) A history of Greek mathematics, vol 2. Oxford University Press, Oxford
22. Hejhal DA (1983) The Selberg trace formula for PSL (2, IR), vol 2. Springer, New York
23. Hildebrand FB (1956) Methods of applied mathematics. Prentice Hall Inc., Englewood Cliffs
24. Ince EL (1956) Ordinary differential equations. Dover Publications Inc., New York
25. Ince EL, Sneddon IN (1987) The solution of ordinary differential equations. Longman Scientific & Technical, Essex
26. Jordan C (1947) Calculus of finite differences, 2nd edn. Chelsea Publishing Company, New York
27. Kamke E (1959) Differentialgleichungen, Lösungenmethoden und Lösungen, I. Gewöhnliche Differentialgleichungen, 6th edn. Akademische Verlagsgesellschaft, Geest & Portig K.-G, Leipzig
28. Kelley W (2003) Book review in Bulletin (new series) of the American Mathematical Society, vol 40, no 2, pp 259–262, S 0273-0979(03)00980-7
29. Kelley WG, Peterson AC (1991) Difference equations, an introduction with applications, 2nd edn. Academic Press, Burlington
30. Lakshmikantham V, Trigante D (2002) Theory of difference equations: numerical methods and applications, 2nd edn. Marcel Dekker Inc., New York
31. Locke P (1973) The Mathematical Gazette, vol 57, no 401, pp 191–193
32. Meschkowski H (1959) Differenzengleichungen. Vandenhoeck & Ruprecht, Göttingen, QA 431.M47
33. Naimark MA (1967) Linear differential operators. Ungar Publishing Co., New York
34. Nikiforov AF, Uvarov VB (1988) Special functions of mathematical physics: a unified introduction with applications. Birkhäuser, Basel
35. Nörlund NE (1954) Vorlesungen über Differenzenrechnung. Chelsea Publishing Company, New York
36. Olver FWJ (ed) NIST Digital Library of Mathematical Functions. http://dlmf.nist.gov/
37. Parke WC, Maximon LC, Closed-form second solution to the differential and difference equations for the confluent hypergeometric function in the degenerate case. International Journal of Difference Equations, in press
38. Stakgold I (1998) Green's functions and boundary value problems, 2nd edn. Wiley, New York
39. Temme NM (1986) Uniform asymptotic expansion for a class of polynomials biorthogonal on the unit circle. Constr Approx 2:369–376
40. Temme NM (2003) Large parameter cases of the Gauss hypergeometric function. J Comput Appl Math 153:441–462
41. Teptin AI (1962) On the sign of the Green's function of a linear difference boundary problem of the second order. Doklady Akademiia nauk SSSR 142:1038–1039
42. Weixlbaumer C (2001) Solutions of difference equations with polynomial coefficients, diploma thesis, RISC Linz, Johannes Kepler Universität, Linz, Austria, Jan 2001
43. Wimp J (1984) Computation with recurrence relations. Pitman Advanced Publishing Program, Boston

Index

© Springer International Publishing Switzerland 2016
L.C. Maximon, *Differential and Difference Equations*,
DOI 10.1007/978-3-319-29736-1